Teacher, Student
One-Stop Internet Resources

Log on to
bookn.msscience.com

ONLINE STUDY TOOLS

- Section Self-Check Quizzes
- Interactive Tutor
- Chapter Review Tests
- Standardized Test Practice
- Vocabulary PuzzleMaker

ONLINE RESEARCH

- WebQuest Projects
- Prescreened Web Links
- Career Links
- Internet Labs

INTERACTIVE ONLINE STUDENT EDITION

- Complete Interactive Student Edition available at mhln.com

FOR TEACHERS

- Teacher Bulletin Board
- Teaching Today—Professional Development

SAFETY SYMBOLS

	HAZARD	EXAMPLES	PRECAUTION	REMEDY
DISPOSAL	Special disposal procedures need to be followed.	certain chemicals, living organisms	Do not dispose of these materials in the sink or trash can.	Dispose of wastes as directed by your teacher.
BIOLOGICAL	Organisms or other biological materials that might be harmful to humans	bacteria, fungi, blood, unpreserved tissues, plant materials	Avoid skin contact with these materials. Wear mask or gloves.	Notify your teacher if you suspect contact with material. Wash hands thoroughly.
EXTREME TEMPERATURE	Objects that can burn skin by being too cold or too hot	boiling liquids, hot plates, dry ice, liquid nitrogen	Use proper protection when handling.	Go to your teacher for first aid.
SHARP OBJECT	Use of tools or glassware that can easily puncture or slice skin	razor blades, pins, scalpels, pointed tools, dissecting probes, broken glass	Practice common-sense behavior and follow guidelines for use of the tool.	Go to your teacher for first aid.
FUME	Possible danger to respiratory tract from fumes	ammonia, acetone, nail polish remover, heated sulfur, moth balls	Make sure there is good ventilation. Never smell fumes directly. Wear a mask.	Leave foul area and notify your teacher immediately.
ELECTRICAL	Possible danger from electrical shock or burn	improper grounding, liquid spills, short circuits, exposed wires	Double-check setup with teacher. Check condition of wires and apparatus.	Do not attempt to fix electrical problems. Notify your teacher immediately.
IRRITANT	Substances that can irritate the skin or mucous membranes of the respiratory tract	pollen, moth balls, steel wool, fiberglass, potassium permanganate	Wear dust mask and gloves. Practice extra care when handling these materials.	Go to your teacher for first aid.
CHEMICAL	Chemicals can react with and destroy tissue and other materials	bleaches such as hydrogen peroxide; acids such as sulfuric acid, hydrochloric acid; bases such as ammonia, sodium hydroxide	Wear goggles, gloves, and an apron.	Immediately flush the affected area with water and notify your teacher.
TOXIC	Substance may be poisonous if touched, inhaled, or swallowed.	mercury, many metal compounds, iodine, poinsettia plant parts	Follow your teacher's instructions.	Always wash hands thoroughly after use. Go to your teacher for first aid.
FLAMMABLE	Flammable chemicals may be ignited by open flame, spark, or exposed heat.	alcohol, kerosene, potassium permanganate	Avoid open flames and heat when using flammable chemicals.	Notify your teacher immediately. Use fire safety equipment if applicable.
OPEN FLAME	Open flame in use, may cause fire.	hair, clothing, paper, synthetic materials	Tie back hair and loose clothing. Follow teacher's instruction on lighting and extinguishing flames.	Notify your teacher immediately. Use fire safety equipment if applicable.

 Eye Safety
Proper eye protection should be worn at all times by anyone performing or observing science activities.

 Clothing Protection
This symbol appears when substances could stain or burn clothing.

 Animal Safety
This symbol appears when safety of animals and students must be ensured.

 Handwashing
After the lab, wash hands with soap and water before removing goggles.

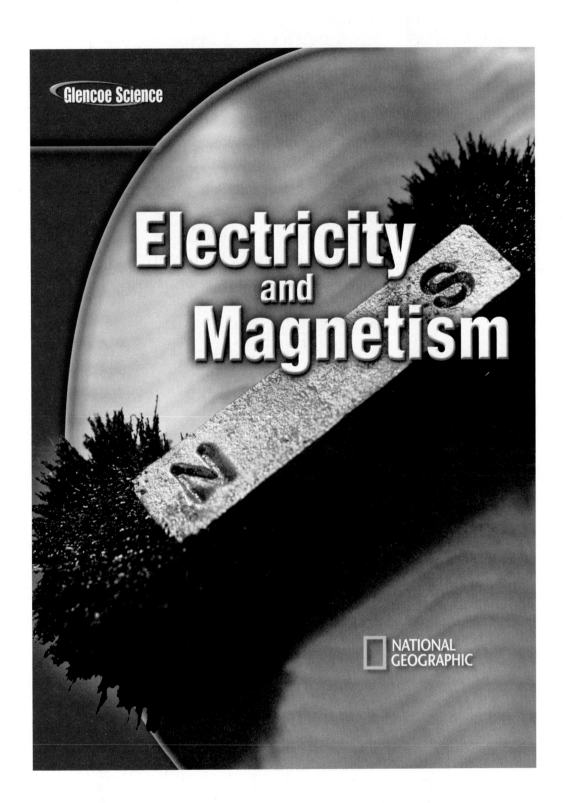

Glencoe Science

Electricity and Magnetism

NATIONAL GEOGRAPHIC

Glencoe

New York, New York   Columbus, Ohio   Chicago, Illinois   Woodland Hills, California

Electricity and Magnetism

Iron filings cluster around the north and south poles of a bar magnet because of magnetic force. Magnetic force is exerted through a magnetic field, which is outlined by the iron filings. This field is caused by negatively charged electrons spinning in the atoms.

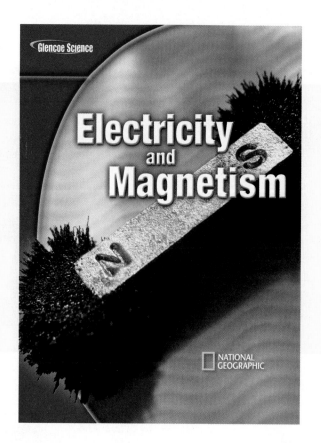

NATIONAL GEOGRAPHIC

 **Glencoe**

The *McGraw-Hill* Companies

Send all inquiries to:
Glencoe/McGraw-Hill
8787 Orion Place
Columbus, OH 43240-4027

ISBN: 978-0-07-877838-4
MHID: 0-07-877838-7

Printed in the United States of America.

2 3 4 5 6 7 8 9 10 027/043 12 11 10 09 08

Authors

NATIONAL GEOGRAPHIC
Education Division
Washington, D.C.

Cathy Ezrailson
Science Department Head
Academy for Science and Health Professions
Conroe, TX

Margaret K. Zorn
Science Writer
Yorktown, VA

Dinah Zike
Educational Consultant
Dinah-Might Activities, Inc.
San Antonio, TX

Series Consultants

CONTENT

Jack Cooper
Ennis High School
Ennis, TX

Carl Zorn, PhD
Staff Scientist
Jefferson Laboratory
Newport News, VA

MATH

Michael Hopper, DEng
Manager of Aircraft Certification
L-3 Communications
Greenville, TX

READING

Barry Barto
Special Education Teacher
John F. Kennedy Elementary
Manistee, MI

Rachel Swaters-Kissinger
Science Teacher
John Boise Middle School
Warsaw, MO

SAFETY

Sandra West, PhD
Department of Biology
Texas State University-San Marcos
San Marcos, TX

ACTIVITY TESTERS

Nerma Coats Henderson
Pickerington Lakeview Jr. High
School
Pickerington, OH

Mary Helen Mariscal-Cholka
William D. Slider Middle School
El Paso, TX

**Science Kit and Boreal
Laboratories**
Tonawanda, NY

Series Reviewers

Deidre Adams
West Vigo Middle School
West Terre Haute, IN

Karen Curry
East Wake Middle School
Raleigh, NC

Anthony J. DiSipio, Jr.
8th Grade Science
Octorana Middle School
Atglen, PA

George Gabb
Great Bridge Middle School
Chesapeake Public Schools
Chesapeake, VA

HOW TO...

Use Your Science Book

Before You Read

- **Chapter Opener** Science is occurring all around you, and the opening photo of each chapter will preview the science you will be learning about. The **Chapter Preview** will give you an idea of what you will be learning about, and you can try the **Launch Lab** to help get your brain headed in the right direction. The **Foldables** exercise is a fun way to keep you organized.

- **Section Opener** Chapters are divided into two to four sections. The **As You Read** in the margin of the first page of each section will let you know what is most important in the section. It is divided into four parts. **What You'll Learn** will tell you the major topics you will be covering. **Why It's Important** will remind you why you are studying this in the first place! The **Review Vocabulary** word is a word you already know, either from your science studies or your prior knowledge. The **New Vocabulary** words are words that you need to learn to understand this section. These words will be in **boldfaced** print and highlighted in the section. Make a note to yourself to recognize these words as you are reading the section.

Glencoe Science

Electricity and Magnetism

NATIONAL GEOGRAPHIC

As You Read

- **Headings** Each section has a title in large red letters, and is further divided into blue titles and small red titles at the beginnings of some paragraphs. To help you study, make an outline of the headings and subheadings.

- **Margins** In the margins of your text, you will find many helpful resources. The **Science Online** exercises and **Integrate** activities help you explore the topics you are studying. **MiniLabs** reinforce the science concepts you have learned.

- **Building Skills** You also will find an **Applying Math** or **Applying Science** activity in each chapter. This gives you extra practice using your new knowledge, and helps prepare you for standardized tests.

- **Student Resources** At the end of the book you will find **Student Resources** to help you throughout your studies. These include **Science, Technology,** and **Math Skill Handbooks,** an **English/Spanish Glossary,** and an **Index.** Also, use your **Foldables** as a resource. It will help you organize information, and review before a test.

- **In Class** Remember, you can always ask your teacher to explain anything you don't understand.

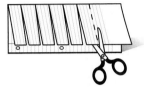

Science Vocabulary Make the following Foldable to help you understand the vocabulary terms in this chapter.

STEP 1 Fold a vertical sheet of notebook paper from side to side.

STEP 2 Cut along every third line of only the top layer to form tabs.

STEP 3 Label each tab with a vocabulary word from the chapter.

Build Vocabulary As you read the chapter, list the vocabulary words on the tabs. As you learn the definitions, write them under the tab for each vocabulary word.

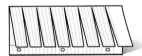

Look For...

FOLDABLES™

At the beginning of every section.

In Lab

Working in the laboratory is one of the best ways to understand the concepts you are studying. Your book will be your guide through your laboratory experiences, and help you begin to think like a scientist. In it, you not only will find the steps necessary to follow the investigations, but you also will find helpful tips to make the most of your time.

- Each lab provides you with a **Real-World Question** to remind you that science is something you use every day, not just in class. This may lead to many more questions about how things happen in your world.

- Remember, experiments do not always produce the result you expect. Scientists have made many discoveries based on investigations with unexpected results. You can try the experiment again to make sure your results were accurate, or perhaps form a new hypothesis to test.

- Keeping a **Science Journal** is how scientists keep accurate records of observations and data. In your journal, you also can write any questions that may arise during your investigation. This is a great method of reminding yourself to find the answers later.

Look For...

- **Launch Labs** start every chapter.
- **MiniLabs** in the margin of each chapter.
- **Two Full-Period Labs** in every chapter.
- **EXTRA Try at Home Labs** at the end of your book.
- the **Web site** with laboratory demonstrations.

Before a Test

Admit it! You don't like to take tests! However, there *are* ways to review that make them less painful. Your book will help you be more successful taking tests if you use the resources provided to you.

- Review all of the **New Vocabulary** words and be sure you understand their definitions.

- Review the notes you've taken on your **Foldables,** in class, and in lab. Write down any question that you still need answered.

- Review the **Summaries** and **Self Check questions** at the end of each section.

- Study the concepts presented in the chapter by reading the **Study Guide** and answering the questions in the **Chapter Review.**

a or b?
?
T or F?

Look For...

- **Reading Checks** and **caption questions** throughout the text.
- the **Summaries** and **Self Check questions** at the end of each section.
- the **Study Guide** and **Review** at the end of each chapter.
- the **Standardized Test Practice** after each chapter.

Let's Get Started

To help you find the information you need quickly, use the Scavenger Hunt below to learn where things are located in Chapter 1.

1. What is the title of this chapter?

2. What will you learn in Section 1?

3. Sometimes you may ask, "Why am I learning this?" State a reason why the concepts from Section 2 are important.

4. What is the main topic presented in Section 2?

5. How many reading checks are in Section 1?

6. What is the Web address where you can find extra information?

7. What is the main heading above the sixth paragraph in Section 2?

8. There is an integration with another subject mentioned in one of the margins of the chapter. What subject is it?

9. List the new vocabulary words presented in Section 2.

10. List the safety symbols presented in the first Lab.

11. Where would you find a Self Check to be sure you understand the section?

12. Suppose you're doing the Self Check and you have a question about concept mapping. Where could you find help?

13. On what pages are the Chapter Study Guide and Chapter Review?

14. Look in the Table of Contents to find out on which page Section 2 of the chapter begins.

15. You complete the Chapter Review to study for your chapter test. Where could you find another quiz for more practice?

Teacher Advisory Board

The Teacher Advisory Board gave the editorial staff and design team feedback on the content and design of the Student Edition. They provided valuable input in the development of the 2008 edition of *Glencoe Science.*

John Gonzales
Challenger Middle School
Tucson, AZ

Rachel Shively
Aptakisic Jr. High School
Buffalo Grove, IL

Roger Pratt
Manistique High School
Manistique, MI

Kirtina Hile
Northmor Jr. High/High School
Galion, OH

Marie Renner
Diley Middle School
Pickerington, OH

Nelson Farrier
Hamlin Middle School
Springfield, OR

Jeff Remington
Palmyra Middle School
Palmyra, PA

Erin Peters
Williamsburg Middle School
Arlington, VA

Rubidel Peoples
Meacham Middle School
Fort Worth, TX

Kristi Ramsey
Navasota Jr. High School
Navasota, TX

Student Advisory Board

The Student Advisory Board gave the editorial staff and design team feedback on the design of the Student Edition. We thank these students for their hard work and creative suggestions in making the 2008 edition of *Glencoe Science* student friendly.

Jack Andrews
Reynoldsburg Jr. High School
Reynoldsburg, OH

Peter Arnold
Hastings Middle School
Upper Arlington, OH

Emily Barbe
Perry Middle School
Worthington, OH

Kirsty Bateman
Hilliard Heritage Middle School
Hilliard, OH

Andre Brown
Spanish Emersion Academy
Columbus, OH

Chris Dundon
Heritage Middle School
Westerville, OH

Ryan Manafee
Monroe Middle School
Columbus, OH

Addison Owen
Davis Middle School
Dublin, OH

Teriana Patrick
Eastmoor Middle School
Columbus, OH

Ashley Ruz
Karrer Middle School
Dublin, OH

The Glencoe middle school science Student Advisory Board taking a timeout at COSI, a science museum in Columbus, Ohio.

Contents

In each chapter, look for these opportunities for review and assessment:

- **Reading Checks**
- **Caption Questions**
- **Section Review**
- **Chapter Study Guide**
- **Chapter Review**
- **Standardized Test Practice**
- **Online practice at bookn.msscience.com**

Get Ready to Read Strategies

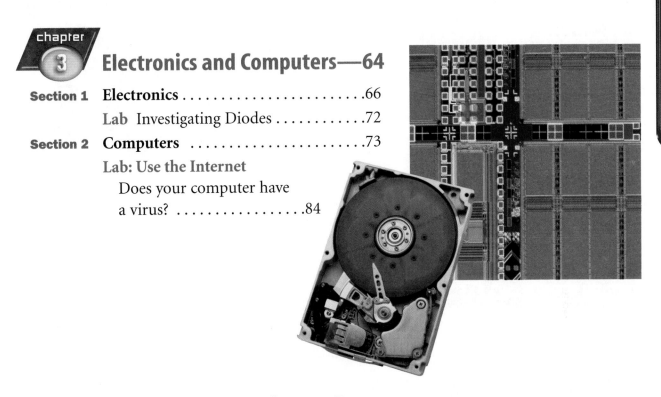

Student Resources

Cross-Curricular Readings/Labs

Content Details

Content Details

Use the Internet Labs

Applying Math

Applying Science

INTEGRATE

Science Online

Standardized Test Practice

Electricity and Magnetism

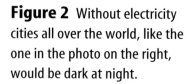

Figure 1 Lightning is a discharge of static electricity.

The brilliant flash of lightning during a thunderstorm might not be new to you. But did you know that it is a discharge of static electricity? It took years and the accumulated work of many scientists to form the foundation of modern ideas about magnetism and electricity.

Writings as early as the first century B.C. show that magnetism was a recognized and studied phenomena. Magnetite, a naturally occurring magnetized rock that attracts iron objects, was available and used to study magnetism by some ancient civilizations. The origins of practical uses for magnetism, such as in compasses, are unknown but they were used centuries before the first writings about magnetism appeared.

Figure 2 Without electricity cities all over the world, like the one in the photo on the right, would be dark at night.

Research and Development

Petrus Peregrinus de Maricourt, a French scientist, published the first documented research study of magnets in 1269. He used magnetite, or lodestone, and a thin, iron rectangle to study the magnetic field generated by the magnetite. Over 300 years later in 1600, William Gilbert, an English physician, published a book called *Of Magnets, Magnetic Bodies, and the Great Magnet of the Earth.* He studied electricity and magnetism and made the analogy that Earth behaves like a giant magnet.

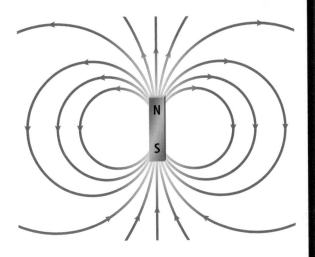

Figure 3 The magnetic field of a bar magnet runs through and around the magnet, from its north pole to its south pole.

The Leyden Jar

In 1746, the Leyden jar was invented by Pieter van Musschenbroek, a Dutch physicist. This provided a cheap and convenient source of electric charges used to study electricity. One form of a Leyden jar is a glass vial partially filled with water that contains a conducting wire capable of storing a large amount of static charge. According to legend, Benjamin Franklin used a Leyden jar when he flew a silk kite during a thunderstorm to show that lightning was an electrical discharge. Franklin's kite was connected by a key on a wet string attached to a Leyden jar. The lightning strike would have caused the Leyden jar to become charged.

Magnetism and Electricity are Related

In 1820, Hans Oersted, a Dutch scientist, discovered that current flowing through a wire deflected a compass needle. This discovery showed a link between electricity and magnetism. Later that year, André Ampère, a French scientist, performed extensive studies on the magnetic fields generated by electric currents and established the laws of magnetic force between electric currents. Michael Faraday, an English scientist, heard about Oersted's work and continued to study the relationship between magnetism and electric current. In 1831, he discovered that moving a magnet near a wire induced a current in the wire. Oersted showed that an electric current creates a magnetic field and Faraday showed that a magnetic field creates an electric current.

Figure 4 Iron filings show the shape of magnetic field lines around various magnets.

Maxwell Ties it All Together

James Maxwell, an English physicist, expanded upon Faraday and Oersted's work and summarized their experimental findings into a set of equations called Maxwell's equations. Maxwell also predicted the existence of electromagnetic waves. He hypothesized that light was an electromagnetic wave. In 1886, Heinrich Hertz, a German physicist, verified the existence of electromagnetic waves when he discovered radio waves.

Electricity and Magnetism Today

These pioneers of science might have had a hard time believing the many ways we use electricity and magnetism today. Our homes contain many devices that use electricity and magnetism. Think about all of the electrical devices that you have used today. Lights, hair dryers, and toasters are just a few of the possibilities that might be on your list. Many of our small appliances contain motors that use both electricity and magnetism to operate. These inventions would not have been possible without the contributions of many scientists.

Physical Science

The study of magnetism and electricity is part of the branch of science known as physical science, or the study of matter and energy. Physical science often is broken into two branches—chemistry and physics. Physics is the study of the interaction between matter and energy. This includes topics such as forces, speed, distance, waves, magnetism, and electricity. Chemistry is the study of the composition, structure, and properties of atoms and matter and the transformations they can undergo.

Figure 5 Huge electric generators like these use the relationship between electricity and magnetism to produce electric current.

The History of Science

The history of magnetism and electricity is an illustration of how science often is the result of the investigations of many people over centuries. Recall that the discovery of the laws that govern magnetism and electricity involved scientists from several countries beginning as early as the first century B.C. This demonstrates the importance of writing and communicating scientific ideas from scientist to scientist and century to century.

Communicating Scientific Ideas

Scientists do not work alone in any field. A good scientist studies historical scientific works as well as contemporary scientific literature. Scientific discoveries must be documented and published so that other scientists can verify results and build upon each other's work. The advancements in electricity and magnetism would have been much slower if each scientist had to rediscover the findings of the scientists before them.

Documenting Scientific Studies

Communication today is almost instantaneous. The telephone, computer, Internet, and fax machine provides us with quick access to people around the world, as well as access to a vast amount of information. There are many ways to document and communicate scientific studies. Journal articles, scientific papers, books, newspaper articles, and web pages are some of the methods that are available today. The documentation of scientific work is as important today as it was centuries ago.

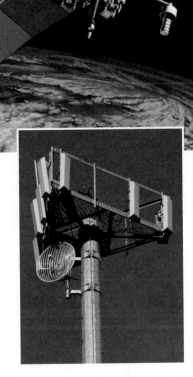

Figure 6 Devices such as satellites (top) and cellular-phone towers (bottom) use electricity and magnetism to enable people to communicate with each other.

You Do It

A modern traveler might use the Global Positioning System, or GPS, instead of a magnetic compass. GPS is often used for navigation instead of a compass because GPS can pinpoint your location to a high degree of accuracy. Research the development of the GPS system. What types of vehicles use GPS devices? Make a list of some of the ways that GPS is used in agriculture, construction, transportation, and research.

The BIG Idea

Electrical energy can be converted into other forms of energy when electric charges flow in a circuit.

Electricity

SECTION 1
Electric Charge

Main Idea Electric charges are positive or negative and exert forces on each other.

SECTION 2
Electric Current

Main Idea A battery produces an electric field in a closed circuit that causes electric charges to flow.

SECTION 3
Electric Circuits

Main Idea Electrical energy can be transferred to devices connected in an electric circuit.

A Blast of Energy

This flash of lightning is an electric spark that releases an enormous amount of electrical energy in an instant. However, in homes and other buildings, electrical energy is released in a controlled way by the flow of electric currents.

Science Journal Write a paragraph describing a lightning flash you have seen. Include information about the weather conditions at the time.

Start-Up Activities

Observing Electric Forces

No computers? No CD players? No video games? Can you imagine life without electricity? Electricity also provides energy that heats and cools homes and produces light. The electrical energy that you use every day is produced by the forces that electric charges exert on each other.

1. Inflate a rubber balloon.

2. Place small bits of paper on your desktop and bring the balloon close to the bits of paper. Record your observations.

3. Charge the balloon by holding it by the knot and rubbing it on your hair or a piece of wool.

4. Bring the balloon close to the bits of paper. Record your observations.

5. Charge two balloons using the procedure in step 3. Hold each balloon by its knot and bring the balloons close to each other. Record your observations.

6. **Think Critically** Compare and contrast the force exerted on the bits of paper by the charged balloon and the force exerted by the two charged balloons on each other.

Electricity Make the following Foldable to help you understand the terms *electric charge, electric current,* and *electric circuit.*

| STEP 1 | Fold the top of a vertical piece of paper down and the bottom up to divide the paper into thirds. |

| STEP 2 | Turn the paper horizontally; unfold and label the three columns as shown. |

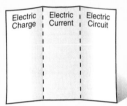

| Electric Charge | Electric Current | Electric Circuit |

Read and Write Before you read the chapter, write a definition of electric charge, electric current, and electric circuit in the appropriate column. As you read the chapter, correct your definition and add additional information about each term.

Science Online Preview this chapter's content and activities at bookn.msscience.com

Get Ready to Read

Make Predictions

1 Learn It! A prediction is an educated guess based on what you already know. One way to predict while reading is to guess what you believe the author will tell you next. As you are reading, each new topic should make sense because it is related to the previous paragraph or passage.

2 Practice It! Read the excerpt below from Section 1. Based on what you have read, make predictions about what you will read in the rest of the lesson. After you read Section 1, go back to your predictions to see if they were correct.

> Predict how the electric force depends on the amount of charge.

> Predict how the electric force would change as charged objects get farther apart.

> Can you predict how the electric force between two electrons changes as the electrons get closer together?

The electric force between two charged objects depends on the distance between them and the **amount of charge** on each object. The electric force between two charged objects gets stronger **as the charges get closer together**. **A positive and a negative charge** are attracted to each other more strongly if they are closer together.

—*from page 11*

3 Apply It! Before you read, skim the questions in the Chapter Review. Choose three questions and predict the answers.

As you read, check the predictions you made to see if they were correct.

Reading Tip

Target Your Reading

Use this to focus on the main ideas as you read the chapter.

1. **Before you read** the chapter, respond to the statements below on your worksheet or on a numbered sheet of paper.
   - Write an **A** if you **agree** with the statement.
   - Write a **D** if you **disagree** with the statement.

2. **After you read** the chapter, look back to this page to see if you've changed your mind about any of the statements.
   - If any of your answers changed, explain why.
   - Change any false statements into true statements.
   - Use your revised statements as a study guide.

Science Online

Print out a worksheet of this page at bookn.msscience.com

Before You Read A or D		Statement	After You Read A or D
	1	Atoms become ions by gaining or losing electrons.	
	2	It is safe to take shelter under a tree during a lightning storm.	
	3	Electric current can follow only one path in a parallel circuit.	
	4	Electrons flow in straight lines through conducting wires.	
	5	Batteries produce electrical energy through nuclear reactions.	
	6	The force between electric charges always is attractive.	
	7	Electrical energy can be transformed into other forms of energy.	
	8	If the voltage in a circuit doesn't change, the current increases if the resistance decreases.	
	9	Electric charges must be touching to exert forces on each other.	

Electric Charge

Electricity

You can't see, smell, or taste electricity, so it might seem mysterious. However, electricity is not so hard to understand when you start by thinking small—very small. All solids, liquids, and gases are made of tiny particles called atoms. Atoms, as shown in **Figure 1,** are made of even smaller particles called protons, neutrons, and electrons. Protons and neutrons are held together tightly in the nucleus at the center of an atom, but electrons swarm around the nucleus in all directions. Protons and electrons have electric charge, but neutrons have no electric charge.

Positive and Negative Charge There are two types of electric charge—positive and negative. Protons have a positive charge, and electrons have a negative charge. The amount of negative charge on an electron is exactly equal to the amount of positive charge on a proton. Because atoms have equal numbers of protons and electrons, the amount of positive charge on all the protons in the nucleus of an atom is balanced by the negative charge on all the electrons moving around the nucleus. Therefore, atoms are electrically neutral, which means they have no overall electric charge.

An atom becomes negatively charged when it gains extra electrons. If an atom loses electrons it becomes positively charged. A positively or negatively charged atom is called an **ion** (I ahn).

Figure 1 An atom is made of positively charged protons (orange), negatively charged electrons (red), and neutrons (blue) with no electric charge.
Identify *where the protons and neutrons are located in an atom.*

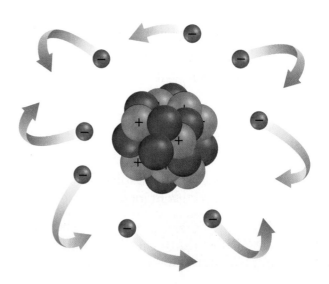

Figure 2 Rubbing can move electrons from one object to another. Hair holds electrons more loosely than the balloon holds them. As a result, electrons are moved from the hair to the balloon when the two make contact. **Infer** *which object has become positively charged and which has become negatively charged.*

Electrons Move in Solids Electrons can move from atom to atom and from object to object. Rubbing is one way that electrons can be transferred. If you have ever taken clinging clothes from a clothes dryer, you have seen what happens when electrons are transferred from one object to another.

Suppose you rub a balloon on your hair. The atoms in your hair hold their electrons more loosely than the atoms on the balloon hold theirs. As a result, electrons are transferred from the atoms in your hair to the atoms on the surface of the balloon, as shown in **Figure 2.** Because your hair loses electrons, it becomes positively charged. The balloon gains electrons and becomes negatively charged. Your hair and the balloon become attracted to one another and make your hair stand on end. This imbalance of electric charge on an object is called a **static charge.** In solids, static charge is due to the transfer of electrons between objects. Protons cannot be removed easily from the nucleus of an atom and usually do not move from one object to another.

Reading Check *How does an object become electrically charged?*

Ions Move in Solutions Sometimes, the movement of charge can be caused by the movement of ions instead of the movement of electrons. Table salt—sodium chloride—is made of sodium ions and chloride ions that are fixed in place and cannot move through the solid. However, when salt is dissolved in water, the sodium and chloride ions break apart and spread out evenly in the water, forming a solution, as shown in **Figure 3.** Now the positive and negative ions are free to move. Solutions containing ions play an important role in enabling different parts of your body to communicate with each other. **Figure 4** shows how a nerve cell uses ions to transmit signals. These signals moving throughout your body enable you to sense, move, and even think.

Figure 3 When table salt (NaCl) dissolves in water, the sodium ions and chloride ions break apart. These ions now are able to carry electric energy.

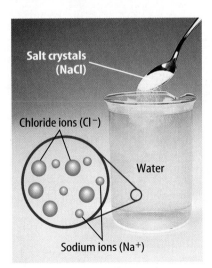

Salt crystals (NaCl)

Chloride ions (Cl$^-$)

Water

Sodium ions (Na$^+$)

NATIONAL GEOGRAPHIC VISUALIZING NERVE IMPULSES

Figure 4

The control and coordination of all your bodily functions involves signals traveling from one part of your body to another through nerve cells. Nerve cells use ions to transmit signals from one nerve cell to another.

A When a nerve cell is not transmitting a signal, it moves positively charged sodium ions (Na^+) outside the membrane of the nerve cell. As a result, the outside of the cell membrane becomes positively charged and the inside becomes negatively charged.

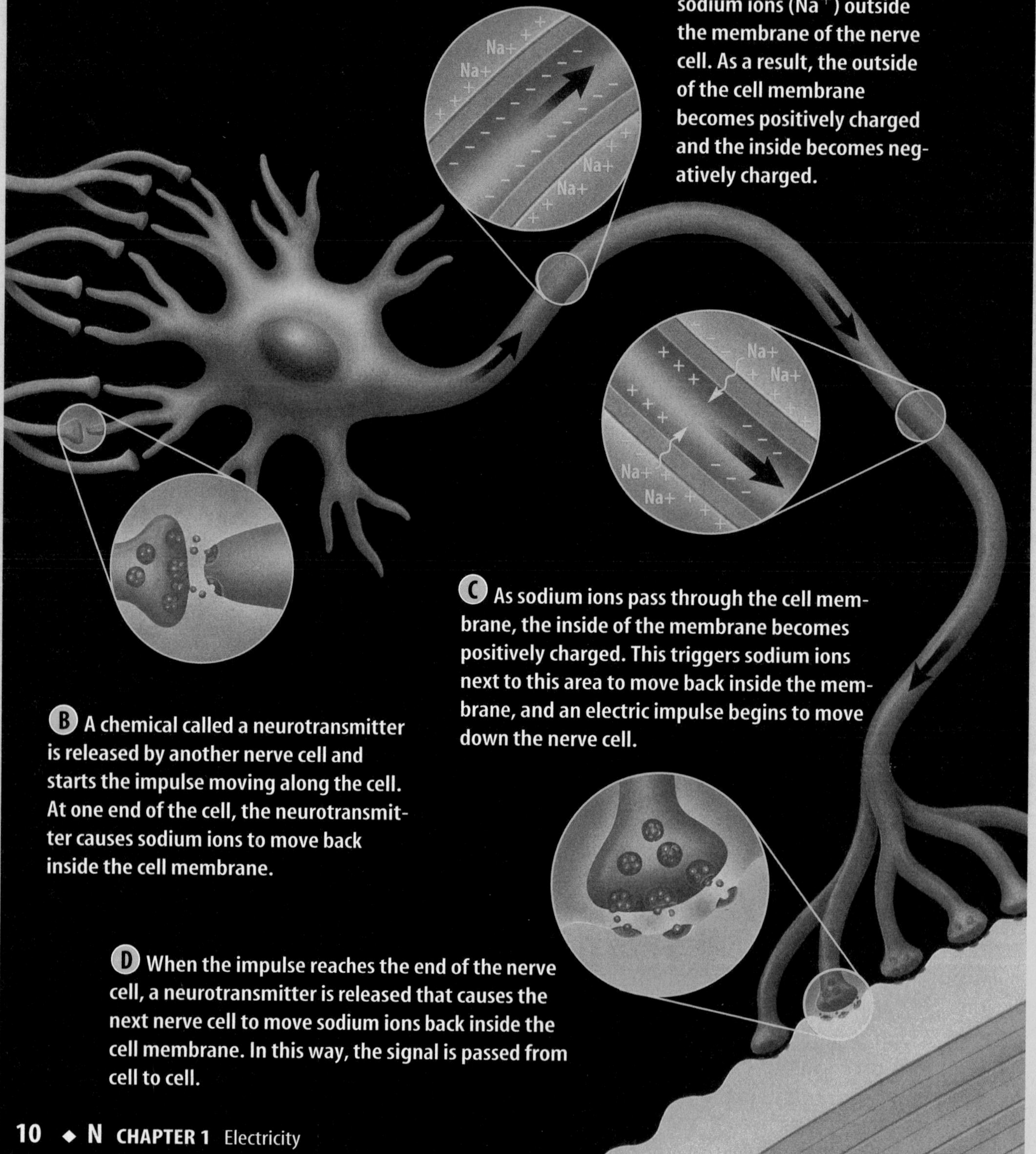

B A chemical called a neurotransmitter is released by another nerve cell and starts the impulse moving along the cell. At one end of the cell, the neurotransmitter causes sodium ions to move back inside the cell membrane.

C As sodium ions pass through the cell membrane, the inside of the membrane becomes positively charged. This triggers sodium ions next to this area to move back inside the membrane, and an electric impulse begins to move down the nerve cell.

D When the impulse reaches the end of the nerve cell, a neurotransmitter is released that causes the next nerve cell to move sodium ions back inside the cell membrane. In this way, the signal is passed from cell to cell.

Unlike charges attract.

Like charges repel. Like charges repel.

Figure 5 A positive charge and a negative charge attract each other. Two positive charges repel each other, as do two negative charges.

Electric Forces

The electrons in an atom swarm around the nucleus. What keeps these electrons close to the nucleus? The positively charged protons in the nucleus exert an attractive electric force on the negatively charged electrons. All charged objects exert an **electric force** on each other. The electric force between two charges can be attractive or repulsive, as shown in **Figure 5.** Objects with the same type of charge repel one another and objects with opposite charges attract one another. This rule is often stated as "like charges repel, and unlike charges attract."

The electric force between two charged objects depends on the distance between them and the amount of charge on each object. The electric force between two electric charges gets stronger as the charges get closer together. A positive and a negative charge are attracted to each other more strongly if they are closer together. Two like charges are pushed away more strongly from each other the closer they are. The electric force between two objects that are charged, such as two balloons that have been rubbed on wool, increases if the amount of charge on at least one of the objects increases.

 **Reading Check** *How does the electric force between two charged objects depend on the distance between them?*

Electric Fields You might have noticed examples of how charged objects don't have to be touching to exert an electric force on each other. For instance, two charged balloons push each other apart even though they are not touching. How are charged objects able to exert forces on each other without touching?

Electric charges exert a force on each other at a distance through an **electric field** that exists around every electric charge. **Figure 6** shows the electric field around a positive and a negative charge. An electric field gets stronger as you get closer to a charge, just as the electric force between two charges becomes greater as the charges get closer together.

Figure 6 The lines with arrowheads represent the electric field around charges. The direction of each arrow is the direction a positive charge would move if it were placed in the field.

The electric field arrows point away from a positive charge.

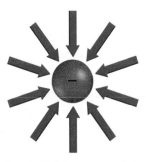

The electric field arrows point toward a negative charge. **Explain** *why the electric field arrows around a negative charge are in the opposite direction of the arrows around a positive charge.*

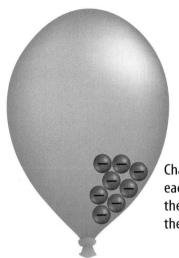

Charges placed on an insulator repel each other but cannot move easily on the surface of the insulator. As a result, the charges remain in one place.

Figure 7 Electric charges move more easily through conductors than through insulators.

The three wires in this electric cable are made of copper, which is a conductor. The wires are covered with plastic insulation that keeps the copper wires from touching each other.

Insulators and Conductors

Rubbing a balloon on your hair transfers electrons from your hair to the balloon. However, only the part of the balloon that was rubbed on your hair becomes charged, because electrons cannot move easily through rubber. As a result, the electrons that were rubbed onto the balloon tend to stay in one place, as shown in **Figure 7.** A material in which electrons cannot move easily from place to place is called an **insulator.** Examples of insulators are plastic, wood, glass, and rubber.

Materials that are **conductors** contain electrons that can move more easily in the material. The electric wire in **Figure 7** is made from a conductor coated with an insulator such as plastic. Electrons move easily in the conductor but do not move easily through the plastic insulation. This prevents electrons from moving through the insulation and causing an electric shock if someone touches the wire.

Metals as Conductors The best conductors are metals such as copper, gold, and aluminum. In a metal atom, a few electrons are not attracted as strongly to the nucleus as the other electrons, and are loosely bound to the atom. When metal atoms form a solid, the metal atoms can move only short distances. However, the electrons that are loosely bound to the atoms can move easily in the solid piece of metal. In an insulator, the electrons are bound tightly in the atoms that make up the insulator and therefore cannot move easily.

Topic: Superconductors
Visit bookn.msscience.com for Web links to information about materials that are superconductors.

Activity Make a table listing five materials that can become superconductors and the critical temperature for each material.

Induced Charge

Has this ever happened to you? You walk across a carpet and as you reach for a metal doorknob, you feel an electric shock. Maybe you even see a spark jump between your fingertip and the doorknob. To find out what happened, look at **Figure 8.**

As you walk, electrons are rubbed off the rug by your shoes. The electrons then spread over the surface of your skin. As you bring your hand close to the doorknob, the electric field around the excess electrons on your hand repels the electrons in the doorknob. Because the doorknob is a good conductor, its electrons move easily away from your hand. The part of the doorknob closest to your hand then becomes positively charged. This separation of positive and negative charges due to an electric field is called an induced charge.

If the electric field between your hand and the knob is strong enough, charge can be pulled from your hand to the doorknob, as shown in **Figure 8.** This rapid movement of excess charge from one place to another is an **electric discharge.** Lightning is an example of an electric discharge. In a storm cloud, air currents sometimes cause the bottom of the cloud to become negatively charged. This negative charge induces a positive charge in the ground below the cloud. A cloud-to-ground lightning stroke occurs when electric charge moves between the cloud and the ground.

Figure 8 A spark that jumps between your fingers and a metal doorknob starts at your feet. **Identify** *another example of an electric discharge.*

As you walk across the floor, you rub electrons from the carpet onto the bottom of your shoes. These electrons then spread out over your skin, including your hands.

As you bring your hand close to the metal doorknob, electrons on the doorknob move as far away from your hand as possible. The part of the doorknob closest to your hand is left with a positive charge.

The attractive electric force between the electrons on your hand and the induced positive charge on the doorknob might be strong enough to pull electrons from your hand to the doorknob. You might see this as a spark and feel a mild electric shock.

Grounding

Lightning is an electric discharge that can cause damage and injury because a lightning bolt releases an extremely large amount of electric energy. Even electric discharges that release small amounts of energy can damage delicate circuitry in devices such as computers. One way to avoid the damage caused by electric discharges is to make the excess charges flow harmlessly into Earth's surface. Earth can be a conductor, and because it is so large, it can absorb an enormous quantity of excess charge.

The process of providing a pathway to drain excess charge into Earth is called grounding. The pathway is usually a conductor such as a wire or a pipe. You might have noticed lightning rods at the top of buildings and towers, as shown in **Figure 9.** These rods are made of metal and are connected to metal cables that conduct electric charge into the ground if the rod is struck by lightning.

Figure 9 A lightning rod can protect a building from being damaged by a lightning strike.

---

section 1 review

Summary

Electric Charges

- There are two types of electric charge—positive charge and negative charge.
- The amount of negative charge on an electron is equal to the amount of positive charge on a proton.
- Objects that are electrically neutral become negatively charged when they gain electrons and positively charged when they lose electrons.

Electric Forces

- Like charges repel and unlike charges attract.
- The force between two charged objects increases as they get closer together.
- A charged object is surrounded by an electric field that exerts a force on other charged objects.

Insulators and Conductors

- Electrons cannot move easily in an insulator but can move easily in a conductor.

Self Check

1. **Explain** why when objects become charged it is electrons that are transferred from one object to another rather than protons.
2. **Compare and contrast** the movement of electric charge in a solution with the transfer of electric charge between solid objects.
3. **Explain** why metals are good conductors.
4. **Compare and contrast** the electric field around a negative charge and the electric field around a positive charge.
5. **Explain** why an electric discharge occurs.
6. **Think Critically** A cat becomes negatively charged when it is brushed. How does the electric charge on the brush compare to the charge on the cat?

Applying Skills

7. **Analyze** You slide out of a car seat and as you touch the metal car door, a spark jumps between your hand and the door. Describe how the spark was formed.

Science online bookn.msscience.com/self_check_quiz

Electric Current

Flow of Charge

An electric discharge, such as a lightning bolt, can release a huge amount of energy in an instant. However, electric lights, refrigerators, TVs, and stereos need a steady source of electrical energy that can be controlled. This source of electrical energy comes from an **electric current,** which is the flow of electric charge. In solids, the flowing charges are electrons. In liquids, the flowing charges are ions, which can be positively or negatively charged. Electric current is measured in units of amperes (A). A model for electric current is flowing water. Water flows downhill because a gravitational force acts on it. Similarly, electrons flow because an electric force acts on them.

A Model for a Simple Circuit How does a flow of water provide energy? If the water is separated from Earth by using a pump, the higher water now has gravitational potential energy, as shown in **Figure 10.** As the water falls and does work on the waterwheel, the water loses potential energy and the waterwheel gains kinetic energy. For the water to flow continuously, it must flow through a closed loop. Electric charges will flow continuously only through a closed conducting loop called a **circuit.**

as you read

What **You'll Learn**

■ **Relate** voltage to the electrical energy carried by an electric current.
■ **Describe** a battery and how it produces an electric current.
■ **Explain** electrical resistance.

Why **It's Important**

Electric current provides a steady source of electrical energy that powers the electric appliances you use every day.

🔍 **Review Vocabulary**
gravitational potential energy: the energy stored in an object due to its position above Earth's surface

New Vocabulary
● electric current   ● voltage
● circuit   ● resistance

Figure 10 The gravitational potential energy of water is increased when a pump raises the water above Earth.

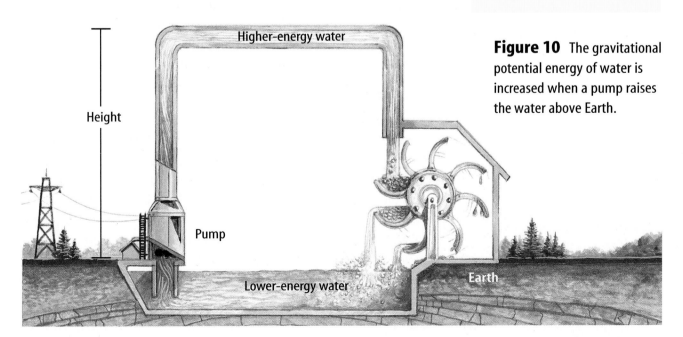

Higher-energy water

Height

Pump

Lower-energy water

Earth

Figure 11 As long as there is a closed path for electrons to follow, electrons can flow in a circuit. They move away from the negative battery terminal and toward the positive terminal.

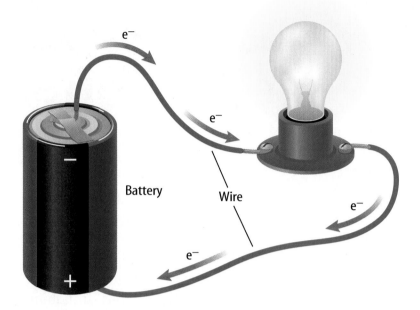

e⁻

Battery   Wire

e⁻

e⁻

e⁻

−

+

Mini LAB

Investigating the Electric Force

Procedure
1. Pour a layer of **salt** on a **plate**.
2. Sparingly sprinkle grains of **pepper** on top of the salt. Do not use too much pepper.
3. Rub a **rubber** or **plastic comb** on an article of **wool clothing**.
4. Slowly drag the comb through the salt and observe.

Analysis
1. How did the salt and pepper react to the comb?
2. Explain why the pepper reacted differently than the salt.

Try at Home

Electric Circuits The simplest electric circuit contains a source of electrical energy, such as a battery, and an electric conductor, such as a wire, connected to the battery. For the simple circuit shown in **Figure 11,** a closed path is formed by wires connected to a lightbulb and to a battery. Electric current flows in the circuit as long as none of the wires, including the glowing filament wire in the lightbulb, is disconnected or broken.

Voltage In a water circuit, a pump increases the gravitational potential energy of the water by raising the water from a lower level to a higher level. In an electric circuit, a battery increases the electrical potential energy of electrons. This electrical potential energy can be transformed into other forms of energy. The **voltage** of a battery is a measure of how much electrical potential energy each electron can gain. As voltage increases, more electrical potential energy is available to be transformed into other forms of energy. Voltage is measured in volts (V).

How a Current Flows You may think that when an electric current flows in a circuit, electrons travel completely around the circuit. Actually, individual electrons move slowly in an electric circuit. When the ends of a wire are connected to a battery, the battery produces an electric field in the wire. The electric field forces electrons to move toward the positive battery terminal. As an electron moves, it collides with other electric charges in the wire and is deflected in a different direction. After each collision, the electron again starts moving toward the positive terminal. A single electron may undergo more than ten trillion collisions each second. As a result, it may take several minutes for an electron in the wire to travel one centimeter.

Batteries A battery supplies energy to an electric circuit. When the positive and negative terminals in a battery are connected in a circuit, the electric potential energy of the electrons in the circuit is increased. As these electrons move toward the positive battery terminal, this electric potential energy is transformed into other forms of energy, just as gravitational potential energy is converted into kinetic energy as water falls.

A battery supplies energy to an electric circuit by converting chemical energy to electric potential energy. For the alkaline battery shown in **Figure 12,** the two terminals are separated by a moist paste. Chemical reactions in the moist paste cause electrons to be transferred to the negative terminal from the atoms in the positive terminal. As a result, the negative terminal becomes negatively charged and the positive terminal becomes positively charged. This produces the electric field in the circuit that causes electrons to move away from the negative terminal and toward the positive terminal.

Battery Life Batteries don't supply energy forever. Maybe you know someone whose car wouldn't start after the lights had been left on overnight. Why do batteries run down? Batteries contain only a limited amount of the chemicals that react to produce chemical energy. These reactions go on as the battery is used and the chemicals are changed into other compounds. Once the original chemicals are used up, the chemical reactions stop and the battery is "dead."

Alkaline Batteries Several chemicals are used to make an alkaline battery. Zinc is a source of electrons at the negative terminal, and manganese dioxide combines with electrons at the positive terminal. The moist paste contains potassium hydroxide that helps transport electrons from the positive terminal to the negative terminal. Research dry-cell batteries and lead-acid batteries. Make a table listing the chemicals used in these batteries and their purpose.

Figure 12 When this alkaline battery is connected in an electric circuit, chemical reactions occur in the moist paste of the battery that move electrons from the positive terminal to the negative terminal.

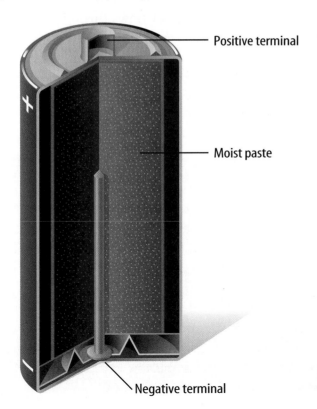

Positive terminal

Moist paste

Negative terminal

The Ohm The unit for electrical resistance was named in honor of the German physicist Georg Simon Ohm (1787–1854). Ohm is credited for discovering the relationship between current flow, voltage, and resistance. Research and find out more about Georg Ohm. Write a brief biography of him to share with the class.

Resistance

Electrons can move much more easily through conductors than through insulators, but even conductors interfere somewhat with the flow of electrons. The measure of how difficult it is for electrons to flow through a material is called **resistance.** The unit of resistance is the ohm (Ω). Insulators generally have much higher resistance than conductors.

As electrons flow through a circuit, they collide with the atoms and other electric charges in the materials that make up the circuit. Look at **Figure 13.** These collisions cause some of the electrons' electrical energy to be converted into thermal energy—heat—and sometimes into light. The amount of electrical energy that is converted into heat and light depends on the resistance of the materials in the circuit.

Buildings Use Copper Wires The amount of electrical energy that is converted into thermal energy increases as the resistance of the wire increases. Copper has low resistance and is one of the best electric conductors. Because copper is a good conductor, less heat is produced as electric current flows in copper wires, compared to wires made of other materials. As a result, copper wire is used in household wiring because the wires usually don't become hot enough to cause fires.

Resistance of Wires The electric resistance of a wire also depends on the length and thickness of the wire, as well as the material it is made from. When water flows through a hose, the water flow decreases as the hose becomes narrower or longer, as shown in **Figure 14** on the next page. The electric resistance of a wire increases as the wire becomes longer or as it becomes narrower.

Figure 13 As electrons flow through a wire, they travel in a zigzag path as they collide with atoms and other electrons. In these collisions, electrical energy is converted into other forms of energy. **Identify** *the other forms of energy that electrical energy is converted into.*

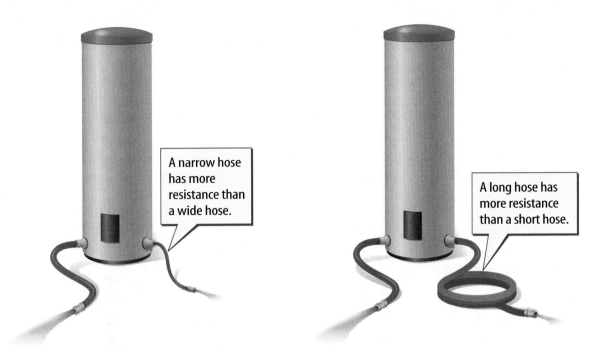

Lightbulb Filaments In a lightbulb, the filament is made of wire so narrow that it has a high resistance. When electric current flows in the filament, it becomes hot enough to emit light. The filament is made of tungsten metal, which has a much higher melting point than most other metals. This keeps the filament from melting at the high temperatures needed to produce light.

Figure 14 The resistance of a hose to the flow of water depends on the diameter and length of the hose used.
Compare and contrast *water flowing in a hose and electric current flowing in a wire.*

section 2 review

Flow of Charge

- Electric current is the flow of electric charges.
- Electric charges will flow continuously only through a closed conducting loop, called a circuit.
- The voltage in a circuit is a measure of the electrical potential energy of the electrons in the circuit.
- A battery supplies energy to an electric circuit by increasing the electric potential energy of electrons in the circuit.

Resistance

- Electric resistance is the measure of how difficult it is for electrons to flow through a material.
- Electric resistance is due to collisions between flowing electrons and the atoms in a material.
- Electric resistance in a circuit converts electrical energy into thermal energy and light.

Self Check

1. **Compare and contrast** an electric discharge with an electric current.
2. **Describe** how a battery causes electrons to move in a circuit.
3. **Describe** how the electric resistance of a wire changes as the wire becomes longer. How does the resistance change as the wire becomes thicker?
4. **Explain** why the electric wires in houses are usually made of copper.
5. **Think Critically** In an electric circuit, where do the electrons come from that flow in the circuit?

Applying Skills

6. **Infer** Find the voltage of various batteries such as a watch battery, a camera battery, a flashlight battery, and an automobile battery. Infer whether the voltage produced by a battery is related to its size.

Electric Circuits

as you read

What You'll Learn

- **Explain** how voltage, current, and resistance are related in an electric circuit.
- **Investigate** the difference between series and parallel circuits.
- **Determine** the electric power used in a circuit.
- **Describe** how to avoid dangerous electric shock.

Why It's Important

Electric circuits control the flow of electric current in all electrical devices.

Review Vocabulary

power: the rate at which energy is transferred; power equals the amount of energy transferred divided by the time over which the transfer occurs

New Vocabulary

- Ohm's law
- series circuit
- parallel circuit
- electric power

Controlling the Current

When you connect a conductor, such as a wire or a lightbulb, between the positive and negative terminals of a battery, electrons flow in the circuit. The amount of current is determined by the voltage supplied by the battery and the resistance of the conductor. To help understand this relationship, imagine a bucket with a hose at the bottom, as shown in **Figure 15.** If the bucket is raised, water will flow out of the hose faster than before. Increasing the height will increase the current.

Voltage and Resistance Think back to the pump and waterwheel in **Figure 10.** Recall that the raised water has energy that is lost when the water falls. Increasing the height from which the water falls increases the energy of the water. Increasing the height of the water is similar to increasing the voltage of the battery. Just as the water current increases when the height of the water increases, the electric current in a circuit increases as voltage increases.

If the diameter of the tube in **Figure 15** is decreased, resistance is greater and the flow of the water decreases. In the same way, as the resistance in an electric circuit increases, the current in the circuit decreases.

Figure 15 Raising the bucket higher increases the potential energy of the water in the bucket. This causes the water to flow out of the hose faster.

Ohm's Law A nineteenth-century German physicist, Georg Simon Ohm, carried out experiments that measured how changing the voltage in a circuit affected the current. He found a simple relationship among voltage, current, and resistance in a circuit that is now known as **Ohm's law.** In equation form, Ohm's law often is written as follows.

Ohm's Law

Voltage (in volts) = **current** (in amperes) × **resistance** (in ohms)
$$V = IR$$

According to Ohm's law, when the voltage in a circuit increases the current increases, just as water flows faster from a bucket that is raised higher. However, if the voltage in the circuit doesn't change, then the current in the circuit decreases when the resistance is increased.

Applying Math **Solving a Simple Equation**

VOLTAGE FROM A WALL OUTLET A lightbulb is plugged into a wall outlet. If the lightbulb has a resistance of 220 Ω and the current in the lightbulb is 0.5 A, what is the voltage provided by the outlet?

Solution

1 *This is what you know:*
- current: $I = 0.5$ A
- resistance: $R = 220$ Ω

2 *This is what you need to find:* voltage: V

3 *This is the procedure you need to use:* Substitute the known values for current and resistance into Ohm's law to calculate the voltage:
$V = IR = (0.5$ A$) (220$ Ω$) = 110$ V

4 *Check your answer:* Divide your answer by the resistance 220 Ω. The result should be the given current 0.5 A.

Practice Problems

1. An electric iron plugged into a wall socket has a resistance of 24 Ω. If the current in the iron is 5.0 A, what is the voltage provided by the wall socket?

2. What is the current in a flashlight bulb with a resistance of 30 Ω if the voltage provided by the flashlight batteries is 3.0 V?

3. What is the resistance of a lightbulb connected to a 110-V wall outlet if the current in the lightbulb is 1.0 A?

 **Science Online**

For more practice, visit bookn.msscience.com/math_practice

Identifying Simple Circuits

Procedure

1. The filament in a lightbulb is a piece of wire. For the bulb to light, an electric current must flow through the filament in a complete circuit. Examine the base of a **flashlight bulb** carefully. Where are the ends of the filament connected to the base?
2. Connect one piece of **wire,** a **battery,** and a flashlight bulb to make the bulb light. (There are four possible ways to do this.)

Analysis

Draw and label a diagram showing the path that is followed by the electrons in your circuit. Explain your diagram.

Series and Parallel Circuits

Circuits control the movement of electric current by providing paths for electrons to follow. For current to flow, the circuit must provide an unbroken path for current to follow. Have you ever been putting up holiday lights and had a string that would not light because a single bulb was missing or had burned out and you couldn't figure out which one it was? Maybe you've noticed that some strings of lights don't go out no matter how many bulbs burn out or are removed. These two strings of holiday lights are examples of the two kinds of basic circuits—series and parallel.

Wired in a Line A **series circuit** is a circuit that has only one path for the electric current to follow, as shown in **Figure 16.** If this path is broken, then the current no longer will flow and all the devices in the circuit stop working. If the entire string of lights went out when only one bulb burned out, then the lights in the string were wired as a series circuit. When the bulb burned out, the filament in the bulb broke and the current path through the entire string was broken.

Reading Check *How many different paths can electric current follow in a series circuit?*

In a series circuit, electrical devices are connected along the same current path. As a result, the current is the same through every device. However, each new device that is added to the circuit decreases the current throughout the circuit. This is because each device has electrical resistance, and in a series circuit, the total resistance to the flow of electrons increases as each additional device is added to the circuit. By Ohm's law, if the voltage doesn't change, the current decreases as the resistance increases.

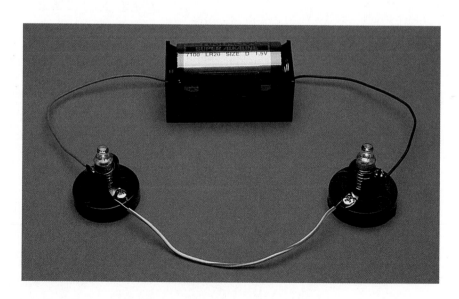

Figure 16 This circuit is an example of a series circuit. A series circuit has only one path for electric current to follow.
Predict *what will happen to the current in this circuit if any of the connecting wires are removed.*

Branched Wiring What if you wanted to watch TV and had to turn on all the lights, a hair dryer, and every other electrical appliance in the house to do so? That's what it would be like if all the electrical appliances in your house were connected in a series circuit.

Instead, houses, schools, and other buildings are wired using parallel circuits. A **parallel circuit** is a circuit that has more than one path for the electric current to follow, as shown in **Figure 17.** The current branches so that electrons flow through each of the paths. If one path is broken, electrons continue to flow through the other paths. Adding or removing additional devices in one branch does not break the current path in the other branches, so the devices on those branches continue to work normally.

In a parallel circuit, the resistance in each branch can be different, depending on the devices in the branch. The lower the resistance is in a branch, the more current flows in the branch. So the current in each branch of a parallel circuit can be different.

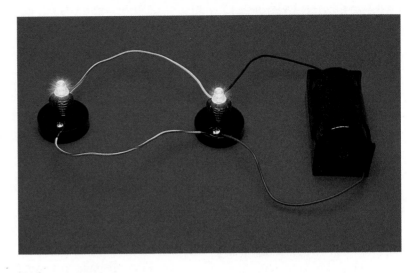

Figure 17 This circuit is an example of a parallel circuit. A parallel circuit has more than one path for electric current to follow. **Predict** *what will happen to the current in the circuit if either of the wires connecting the two lightbulbs is removed.*

Protecting Electric Circuits

In a parallel circuit, the current that flows out of the battery or electric outlet increases as more devices are added to the circuit. As the current through the circuit increases, the wires heat up.

To keep the wire from becoming hot enough to cause a fire, the circuits in houses and other buildings have fuses or circuit breakers like those shown in **Figure 18** that limit the amount of current in the wiring. When the current becomes larger than 15 A or 20 A, a piece of metal in the fuse melts or a switch in the circuit breaker opens, stopping the current. The cause of the overload can then be removed, and the circuit can be used again by replacing the fuse or resetting the circuit breaker.

Fuse

In some buildings, each circuit is connected to a fuse. The fuses are usually located in a fuse box.

Figure 18 You might have fuses in your home that prevent electric wires from overheating.

Wire

A fuse contains a piece of wire that melts and breaks when the current flowing through the fuse becomes too large.

Table 1 Power Used by Common Appliances	
Appliance	Power (W)
Computer	350
Color TV	200
Stereo	250
Refrigerator	450
Microwave	700–1,500
Hair dryer	1,000

Electric Power

When you use an appliance such as a toaster or a hair dryer, electrical energy is converted into other forms of energy. The rate at which electrical energy is converted into other forms of energy is **electric power.** In an electric appliance or in any electric circuit, the electric power that is used can be calculated from the electric power equation.

Electric Power Equation

Power (in watts) = **current** (in amperes) × **voltage** (in volts)

$$P = IV$$

The electric power is equal to the voltage provided to the appliance times the current that flows into the appliance. In the electric power equation, the SI unit of power is the watt. **Table 1** lists the electric power used by some common appliances.

Applying Math — Solving a Simple Equation

ELECTRIC POWER USED BY A LIGHTBULB A lightbulb is plugged into a 110-V wall outlet. How much electric power does the lightbulb use if the current in the bulb is 0.55 A?

Solution

1 *This is what you know:*
- voltage: $V = 110$ V
- current: $I = 0.55$ A

2 *This is what you need to find:*

power: P

3 *This is the procedure you need to use:*

To calculate electric power, substitute the known values for voltage and current into the equation for electric power:
$P = IV = (0.55$ A$)(110$ V$) = 60$ W

4 *Check your answer:*

Divide your answer by the current 0.55 A. The result should be the given voltage 110 V.

Practice Problems

1. The batteries in a portable CD player provide 6.0 V. If the current in the CD player is 0.5 A, how much power does the CD player use?

2. What is the current in a toaster if the toaster uses 1,100 W of power when plugged into a 110-V wall outlet?

3. An electric clothes dryer uses 4,400 W of electric power. If the current in the dryer is 20.0 A, what is the voltage?

For more practice, visit bookn.msscience.com/math_practice

Cost of Electric Energy Power is the rate at which energy is used, or the amount of energy that is used per second. When you use a hair dryer, the amount of electrical energy that is used depends on the power of the hair dryer and the amount of time you use it. If you used it for 5 min yesterday and 10 min today, you used twice as much energy today as yesterday.

Using electrical energy costs money. Electric companies generate electrical energy and sell it in units of kilowatt-hours to homes, schools, and businesses. One kilowatt-hour, kWh, is an amount of electrical energy equal to using 1 kW of power continuously for 1 h. This would be the amount of energy needed to light ten 100-W lightbulbs for 1 h, or one 100-W lightbulb for 10 h.

Reading Check *What does kWh stand for and what does it measure?*

An electric company usually charges its customers for the number of kilowatt-hours they use every month. The number of kilowatt-hours used in a building such as a house or a school is measured by an electric meter, which usually is attached to the outside of the building, as shown in **Figure 19.**

Figure 19 Electric meters measure the amount of electrical energy used in kilowatt-hours. **Identify** *the electric meter attached to your house.*

Electrical Safety

Have you ever had a mild electric shock? You probably felt only a mild tingling sensation, but electricity can have much more dangerous effects. In 1997, electric shocks killed an estimated 490 people in the United States. **Table 2** lists a few safety tips to help prevent electrical accidents.

Table 2 Preventing Electric Shock
Never use appliances with frayed or damaged electric cords.
Unplug appliances before working on them, such as when prying toast out of a jammed toaster.
Avoid all water when using plugged-in appliances.
Never touch power lines with anything, including kite string and ladders.
Always respect warning signs and labels.

Science Online

Topic: Cost of Electrical Energy
Visit bookn.msscience.com for Web links to information about the cost of electrical energy in various parts of the world.

Activity Make a bar graph showing the cost of electrical energy for several countries on different continents.

Current's Effects The scale below shows how the effect of electric current on the human body depends on the amount of current that flows into the body.

0.0005 A	Tingle
0.001 A	Pain threshold
0.01 A	Inability to let go
0.025 A	
0.05 A	Difficulty breathing
0.10 A	
0.25 A	
0.50 A	Heart failure
1.00 A	

Electric Shock You experience an electric shock when an electric current enters your body. In some ways your body is like a piece of insulated wire. The fluids inside your body are good conductors of current. The electrical resistance of dry skin is much higher. Skin insulates the body like the plastic insulation around a copper wire. Your skin helps keep electric current from entering your body.

A current can enter your body when you accidentally become part of an electric circuit. Whether you receive a deadly shock depends on the amount of current that flows into your body. The current that flows through the wires connected to a 60-W light-bulb is about 0.5 A. This amount of current entering your body could be deadly. Even a current as small as 0.001 A can be painful.

Lightning Safety On average, more people are killed every year by lightning in the United States than by hurricanes or tornadoes. Most lightning deaths and injuries occur outdoors. If you are outside and can see lightning or hear thunder, take shelter indoors immediately. If you cannot go indoors, you should take these precautions: avoid high places and open fields; stay away from tall objects such as trees, flag poles, or light towers; and avoid objects that conduct current such as bodies of water, metal fences, picnic shelters, and metal bleachers.

section 3 review

Summary

Electric Circuits

- In an electric circuit, voltage, resistance, and current are related. According to Ohm's law, this relationship can be written as $V = IR$.

- A series circuit has only one path for electric current to follow.

- A parallel circuit has more than one path for current to follow.

Electric Power and Energy

- The electric power used by an appliance is the rate at which the appliance converts electrical energy to other forms of energy.

- The electric power used by an appliance can be calculated using the equation $P = IV$.

- The electrical energy used by an appliance depends on the power of the appliance and the length of time it is used. Electrical energy usually is measured in kWh.

Self Check

1. **Compare** the current in two lightbulbs wired in a series circuit.

2. **Describe** how the current in a circuit changes if the resistance increases and the voltage remains constant.

3. **Explain** why buildings are wired using parallel circuits rather than series circuits.

4. **Identify** what determines the damage caused to the human body by an electric shock.

5. **Think Critically** What determines whether a 100-W lightbulb costs more to use than a 1,200-W hair dryer costs to use?

Applying Math

6. **Calculate Energy** A typical household uses 1,000 kWh of electrical energy every month. If a power company supplies electrical energy to 1,000 households, how much electrical energy must it supply every year?

Current in a Parallel Circuit

The brightness of a lightbulb increases as the current in the bulb increases. In this lab you'll use the brightness of a lightbulb to compare the amount of current that flows in parallel circuits.

▶ Real-World Question

How does connecting devices in parallel affect the electric current in a circuit?

Goal
- **Observe** how the current in a parallel circuit changes as more devices are added.

Materials
1.5-V lightbulbs (4)   battery holders (2)
1.5-V batteries (2)   minibulb sockets (4)
10-cm-long pieces of
   insulated wire (8)

Safety Precautions

▶ Procedure

1. Connect one lightbulb to the battery in a complete circuit. After you've made the bulb light, disconnect the bulb from the battery to keep the battery from running down. This circuit will be the brightness tester.

2. Make a parallel circuit by connecting two bulbs as shown in the diagram. Reconnect the bulb in the brightness tester and compare its brightness with the brightness of the two bulbs in the parallel circuit. Record your observations.

3. Add another bulb to the parallel circuit as shown in the figure. How does the brightness of the bulbs change?

4. Disconnect one bulb in the parallel circuit. Record your observations.

▶ Conclude and Apply

1. **Describe** how the brightness of each bulb depends on the number of bulbs in the circuit.

2. **Infer** how the current in each bulb depends on the number of bulbs in the circuit.

𝒞ommunicating
Your Data

Compare your conclusions with those of other students in your class. **For more help, refer to the** Science Skill Handbook.

A Model for Voltage and Current

Goal

■ **Model** the flow of current in a simple circuit.

Materials

plastic funnel
rubber or plastic tubing
   of different diameters
   (1 m each)
meterstick
ring stand with ring
stopwatch
*clock displaying seconds
hose clamp
*binder clip
500-mL beakers (2)
*Alternate materials

Safety Precautions

◉ Real-World Question

The flow of electrons in an electric circuit is something like the flow of water in a tube connected to a water tank. By raising or lowering the height of the tank, you can increase or decrease the potential energy of the water. How does the flow of water in a tube depend on the diameter of the tube and the height the water falls?

◉ Procedure

1. **Design** a data table in which to record your data. It should be similar to the table below.

2. Connect the tubing to the bottom of the funnel and place the funnel in the ring of the ring stand.

3. **Measure** the inside diameter of the rubber tubing. Record your data.

4. Place a 500-mL beaker at the bottom of the ring stand and lower the ring so the open end of the tubing is in the beaker.

5. Use the meterstick to measure the height from the top of the funnel to the bottom of the ring stand.

Flow Rate Data				
Trial	Height (cm)	Diameter (mm)	Time (s)	Flow Rate (mL/s)
1				
2				
3		Do not write in this book.		
4				

6. Working with a classmate, pour water into the funnel fast enough to keep the funnel full but not overflowing. Measure and record the time needed for 100 mL of water to flow into the beaker. Use the hose clamp to start and stop the flow of water.

7. Connect tubing with a different diameter to the funnel and repeat steps 2 through 6.

8. Reconnect the original piece of tubing and repeat steps 4 through 6 for several lower positions of the funnel, lowering the height by 10 cm each time.

Analyze Your Data

1. **Calculate** the rate of flow for each trial by dividing 100 mL by the time measured for 100 mL of water to flow into the beaker.

2. **Make a graph** that shows how the rate of flow depends on the funnel height.

Conclude and Apply

1. **Infer** from your graph how the rate of flow depends on the height of the funnel.

2. **Explain** how the rate of flow depends on the diameter of the tubing. Is this what you expected to happen?

3. **Identify** which of the variables you changed in your trials that corresponds to the voltage in a circuit.

4. **Identify** which of the variables you changed in your trials that corresponds to the resistance in a circuit.

5. **Infer** from your results how the current in a circuit would depend on the voltage.

6. **Infer** from your results how the current in a circuit would depend on the resistance in the circuit.

Communicating Your Data

Share your graph with other students in your class. Did other students draw the same conclusions as you? **For more help, refer to the** Science Skill Handbook.

Fire in the Forest

Plant life returns after a forest fire in Yellowstone National Park.

Fires started by lightning may not be all bad

When lightning strikes a tree, the intense heat of the lightning bolt can set the tree on fire. The fire then can spread to other trees in the forest. Though lightning is responsible for only about ten percent of forest fires, it causes about one-half of all fire damage. For example, in 2000, fires set by lightning raged in 12 states at the same time, burning a total area roughly the size of the state of Massachusetts.

Fires sparked by lightning often start in remote, difficult-to-reach areas, such as national parks and range lands. Burning undetected for days, these fires can spread out of control. In addition to threatening lives, the fires can destroy millions of dollars worth of homes and property. Smoke from forest fires also can have harmful effects on people, especially for those with preexisting conditions, such as asthma.

People aren't the only victims of forest fires. The fires kill animals as well. Those who survive the blaze often perish because their habitats have been destroyed. Monster blazes spew carbon dioxide and other gases into the atmosphere. Some of these gases may contribute to the greenhouse effect that warms the planet. Moreover, fires cause soil erosion and loss of water reserves.

But fires caused by lightning also have some positive effects. In old, dense forests, trees often become diseased and insect-ridden. By removing these unhealthy trees, fires allow healthy trees greater access to water and nutrients. Fires also clear away a forest's dead trees, underbrush, and needles, providing space for new vegetation. Nutrients are returned to the ground as dead organic matter decays, but it can take a century for dead logs to rot completely. A fire enables nutrients to be returned to soil much more quickly. Also, the removal of these highly combustible materials prevents more widespread fires from occurring.

Research Find out more about the job of putting out forest fires. What training is needed? What gear do firefighters wear? Why would people risk their lives to save a forest? Use the media center at your school to learn more about forest firefighters and their careers.

Science Online

For more information, visit bookn.msscience.com/time

Section 1 Electric Charge

1. The two types of electric charge are positive and negative. Like charges repel and unlike charges attract.

2. An object becomes negatively charged if it gains electrons and positively charged if it loses electrons.

3. Electrically charged objects have an electric field surrounding them and exert electric forces on one another.

4. Electrons can move easily in conductors, but not so easily in insulators.

Section 2 Electric Current

1. Electric current is the flow of charges—usually either electrons or ions.

2. The energy carried by the current in a circuit increases as the voltage in the circuit increases.

3. In a battery, chemical reactions provide the energy that causes electrons to flow in a circuit.

4. As electrons flow in a circuit, some of their electrical energy is lost due to resistance in the circuit.

Section 3 Electric Circuits

1. In an electric circuit, the voltage, current, and resistance are related by Ohm's law.

2. The two basic kinds of electric circuits are parallel circuits and series circuits.

3. The rate at which electric devices use electrical energy is the electric power used by the device.

Visualizing Main Ideas

Copy and complete the following concept map about electricity.

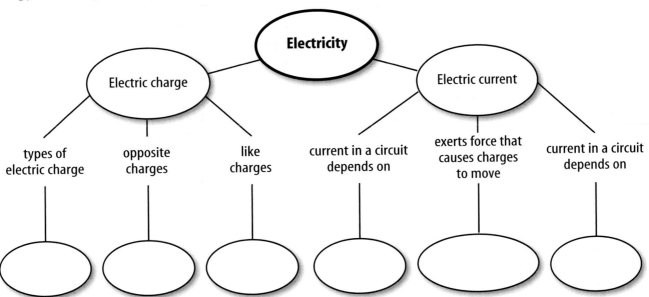

Using Vocabulary

circuit p. 15
conductor p. 12
electric current p. 15
electric discharge p. 13
electric field p. 11
electric force p. 11
electric power p. 24
insulator p. 12

ion p. 8
Ohm's law p. 21
parallel circuit p. 23
resistance p. 18
series circuit p. 22
static charge p. 9
voltage p. 16

Answer the following questions using complete sentences.

1. What is the term for the flow of electric charge?

2. What is the relationship among voltage, current, and resistance in a circuit?

3. In what type of material do electrons move easily?

4. What is the name for the unbroken path that current follows?

5. What is an excess of electric charge on an object?

6. What is an atom that has gained or lost electrons called?

7. Which type of circuit has more than one path for electrons to follow?

8. What is the rapid movement of excess charge known as?

Checking Concepts

Choose the word or phrase that best answers the question.

9. Which of the following describes an object that is positively charged?
   A) has more neutrons than protons
   B) has more protons than electrons
   C) has more electrons than protons
   D) has more electrons than neutrons

10. Which of the following is true about the electric field around an electric charge?
    A) It exerts a force on other charges.
    B) It increases the resistance of the charge.
    C) It increases farther from the charge.
    D) It produces protons.

11. What is the force between two electrons?
    A) frictional   C) attractive
    B) neutral   D) repulsive

12. What property of a wire increases when it is made thinner?
    A) resistance
    B) voltage
    C) current
    D) static charge

13. What property does Earth have that enables grounding to drain static charges?
    A) It has a high static charge.
    B) It has a high resistance.
    C) It is a large conductor.
    D) It is like a battery.

Use the graph below to answer question 14.

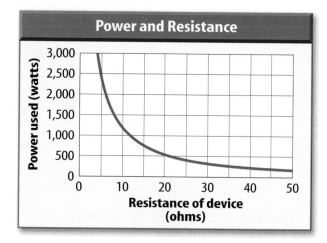

14. How does the resistance change if the power decreases from 2,500 W to 500 W?
    A) It increases four times.
    B) It decreases four times.
    C) It doubles.
    D) It doesn't change.

Science Online   bookn.msscience.com/vocabulary_puzzlemaker

Thinking Critically

15. Determine A metal wire is made thinner. How would you change the length of the wire to keep the electric resistance of the wire from changing?

The tables below show how the voltage and current vary in a portable radio and a portable CD player. Use these tables to answer questions 16 through 19.

Portable Radio		Portable CD Player	
Voltage (V)	Current (A)	Voltage (V)	Current (A)
2.0	1.0	2.0	0.5
4.0	2.0	4.0	1.0
6.0	3.0	6.0	1.5

16. Make a graph with current plotted on the horizontal axis and voltage plotted on the vertical axis. Plot the data in the above tables for both devices on your graph.

17. Identify from your graph which line is more horizontal—the line for the portable radio or the line for the portable CD player.

18. Calculate the electric resistance using Ohm's law for each value of the current and voltage in the tables above. What is the resistance of each device?

19. Determine For which device is the line plotted on your graph more horizontal—the device with higher or lower resistance?

20. Explain why a balloon that has a static electric charge will stick to a wall.

21. Describe how you can tell whether the type of charge on two charged objects is the same or different.

22. Infer Measurements show that Earth is surrounded by a weak electric field. If the direction of this field points toward Earth, what is the type of charge on Earth's surface?

Performance Activities

23. Design a board game about a series or parallel circuit. The rules of the game could be based on opening or closing the circuit, adding more devices to the circuit, blowing fuses or circuit breakers, replacing fuses, or resetting circuit breakers.

Applying Math

24. Calculate Resistance A toaster is plugged into a 110-V outlet. What is the resistance of the toaster if the current in the toaster is 10 A?

25. Calculate Current A hair dryer uses 1,000 W when it is plugged into a 110-V outlet. What is the current in the hair dryer?

26. Calculate Voltage A lightbulb with a resistance of 30 Ω is connected to a battery. If the current in the lightbulb is 0.10 A, what is the voltage of the battery?

Use the table below to answer question 27.

Average Standby Power Used	
Appliance	Power (W)
Computer	7.0
VCR	6.0
TV	5.0

27. Calculate Cost The table above shows the power used by several appliances when they are turned off. Calculate the cost of the electrical energy used by each appliance in a month if the cost of electrical energy is $0.08/kWh, and each appliance is in standby mode for 600 h each month.

Part 1 | Multiple Choice

Record your answers on the answer sheet provided by your teacher or on a sheet of paper.

1. What happens when two materials are charged by rubbing against each other?
   A. both lose electrons
   B. both gain electrons
   C. one loses electrons
   D. no movement of electrons

Use the table below to answer questions 2–4.

Power Ratings of Some Appliances	
Appliance	**Power (W)**
Computer	350
Color TV	200
Stereo	250
Toaster	1,100
Microwave	900
Hair dryer	1,000

2. Which appliance will use the most energy if it is run for 15 minutes?
   A. microwave   C. stereo
   B. computer   D. color TV

3. What is the current in the hair dryer if it is plugged into a 110-V outlet?
   A. 110 A   C. 9 A
   B. 130,000 A   D. 1,100 A

4. Suppose using 1,000 W for 1 h costs $0.10. How much would it cost to run the color TV for 8 hours?
   A. $1.00   C. $1.60
   B. $10.00   D. $0.16

5. How does the current in a circuit change if the voltage is doubled and the resistance remains unchanged?
   A. no change   C. doubles
   B. triples   D. reduced by half

6. Which statement does NOT describe how electric changes affect each other?
   A. positive and negative charges attract
   B. positive and negative charges repel
   C. two positive charges repel
   D. two negative charges repel

Use the illustration below to answer questions 7 and 8.

7. What is the device on the chimney called?
   A. circuit breaker   C. fuse
   B. lightning rod   D. circuit

8. What is the device designed to do?
   A. stop electricity from flowing
   B. repel an electric charge
   C. turn the chimney into an insulator
   D. to provide grounding for the house

9. Which of the following is a material through which charge cannot move easily?
   A. conductor   C. wire
   B. circuit   D. insulator

10. What property of a wire increases when it is made longer?
    A. charge   C. voltage
    B. resistance   D. current

11. Which of the following materials are good insulators?
    A. copper and gold
    B. wood and glass
    C. gold and aluminum
    D. plastic and copper

Part 2 | Short Response/Grid In

Record your answers on the answer sheet provided by your teacher or on a sheet of paper.

Use the illustration below to answer questions 12 and 13.

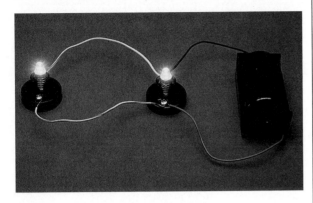

12. In this circuit, if one lightbulb is unscrewed, what happens to the current in the other lightbulb? Explain.

13. In this circuit, is the resistance and the current in each branch of the circuit always the same? Explain.

14. A 1,100-W toaster may be used for five minutes each day. A 400-W refrigerator runs all the time. Which appliance uses more electrical energy? Explain.

15. How much current does a 75-W bulb require in a 100-V circuit?

16. A series circuit containing mini-lightbulbs is opened and some of the lightbulbs are removed. What happens when the circuit is closed?

17. Suppose you plug an electric heater into the wall outlet. As soon as you turn it on, all the lights in the room go out. Explain what must have happened.

18. Explain why copper wires used in appliances or electric circuits are covered with plastic or rubber.

Part 3 | Open Ended

Record your answers on a sheet of paper.

19. Why is it dangerous to use a fuse that is rated 30 A in a circuit calling for a 15-A fuse?

Use the illustration below to answer question 20.

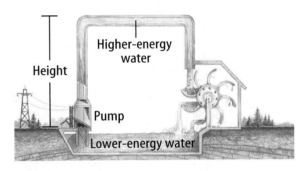

20. Compare the water pump in the water circuit above with the battery in an electric circuit.

21. Explain what causes the lightning that is associated with a thunderstorm.

22. Explain why two charged balloons push each other apart even if they are not touching.

23. Explain what can happen when you rub your feet on a carpet and then touch a metal doorknob.

24. Why does the fact that tungsten wire has a high melting point make it useful in the filaments of lightbulbs?

Test-Taking Tip

Recall Experiences Recall any hands-on experience as you read the question. Base your answer on the information given on the test.

Question 23 Recall from your personal experience the jolt you feel when you touch a doorknob after walking across a carpet.

Magnetism

The BIG Idea

Magnets exert forces on other magnets and on moving charges.

SECTION 1
What is magnetism?
Main Idea Moving electric charges produce magnetic fields.

SECTION 2
Electricity and Magnetism
Main Idea Magnetic fields can produce electric currents.

Magnetic Suspension

This experimental train can travel at speeds as high as 500 km/h— without even touching the track! It uses magnetic levitation, or maglev, to reach these high speeds. Magnetic forces lift the train above the track, and propel it forward at high speeds.

Science Journal List three ways that you have seen magnets used.

Start-Up Activities

Magnetic Forces

A maglev is moved along at high speeds by magnetic forces. How can a magnet get something to move? The following lab will demonstrate how a magnet is able to exert forces.

1. Place two bar magnets on opposite ends of a sheet of paper.

2. Slowly slide one magnet toward the other until it moves. Measure the distance between the magnets.

3. Turn one magnet around 180°. Repeat Step 2. Then turn the other magnet and repeat Step 2 again.

4. Repeat Step 2 with one magnet perpendicular to the other, in a T shape.

5. **Think Critically** In your Science Journal, record your results. In each case, how close did the magnets have to be to affect each other? Did the magnets move together or apart? How did the forces exerted by the magnets change as the magnets were moved closer together? Explain.

Preview this chapter's content and activities at bookn.msscience.com

Magnetic Forces and Fields Make the following Foldable to help you see how magnetic forces and magnetic fields are similar and different.

STEP 1 Draw a mark at the midpoint of a vertical sheet of paper along the side edge.

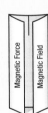

STEP 2 Turn the paper horizontally and **fold** the outside edges in to touch at the midpoint mark.

STEP 3 Label the flaps *Magnetic Force* and *Magnetic Field.*

Compare and Contrast As you read the chapter, write information about each topic on the inside of the appropriate flap. After you read the chapter, compare and contrast the terms *magnetic force* and *magnetic field.* Write your observations under the flaps.

Get Ready to Read

Identify Cause and Effect

① Learn It! A *cause* is the reason something happens. The result of what happens is called an *effect*. Learning to identify causes and effects helps you understand why things happen. By using graphic organizers, you can sort and analyze causes and effects as you read.

② Practice It! Read the following paragraph. Then use the graphic organizer below to show what happened when the Sun ejects charged particles toward Earth.

> Sometimes the Sun ejects a large number of charged particles all at once. Most of these charged particles are deflected by Earth's magnetosphere. However, some of the ejected particles from the Sun produce other charged particles in Earth's outer atmosphere. These charged particles spiral along Earth's magnetic field lines toward Earth's magnetic poles. There they collide with atoms in the atmosphere. These collisions cause the atoms to emit light.
>
> —*from page 49*

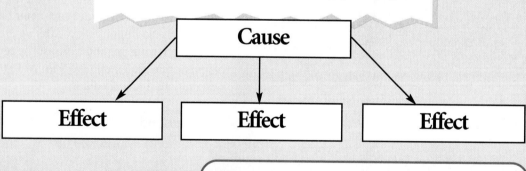

③ Apply It! As you read the chapter, be aware of causes and effects of charged particles moving in a magnetic field. Find three causes and their effects.

Target Your Reading

Use this to focus on the main ideas as you read the chapter.

1 **Before you read** the chapter, respond to the statements below on your worksheet or on a numbered sheet of paper.
- Write an **A** if you **agree** with the statement.
- Write a **D** if you **disagree** with the statement.

2 **After you read** the chapter, look back to this page to see if you've changed your mind about any of the statements.
- If any of your answers changed, explain why.
- Change any false statements into true statements.
- Use your revised statements as a study guide.

Reading Tip

Graphic organizers such as the Cause-Effect organizer help you organize what you are reading so you can remember it later.

Science Online

Print out a worksheet of this page at bookn.msscience.com

Before You Read A or D	Statement	After You Read A or D
	1 Opposite poles of magnets attract each other.	
	2 An electric motor converts electrical energy into kinetic energy.	
	3 Earth's magnetic field has not changed since the Earth formed.	
	4 Magnetic fields get stronger as you move away from the magnet's poles.	
	5 A wire carrying electric current is surrounded by a magnetic field.	
	6 An electromagnet is wire wrapped around a magnet.	
	7 Magnetic fields have no effect on moving electric charges.	
	8 Earth's magnetic field affects only Earth's surface .	
	9 Magnetic fields are produced by moving masses.	
	10 Transformers convert kinetic energy to electrical energy.	

What is magnetism?

as you read

What You'll Learn

- **Describe** the behavior of magnets.
- **Relate** the behavior of magnets to magnetic fields.
- **Explain** why some materials are magnetic.

Why It's Important

Magnetism is one of the basic forces of nature.

⊙ Review Vocabulary

compass: a device which uses a magnetic needle that can turn freely to determine direction

New Vocabulary

- magnetic field
- magnetic domain
- magnetosphere

Early Uses

Do you use magnets to attach papers to a metal surface such as a refrigerator? Have you ever wondered why magnets and some metals attract? Thousands of years ago, people noticed that a mineral called magnetite attracted other pieces of magnetite and bits of iron. They discovered that when they rubbed small pieces of iron with magnetite, the iron began to act like magnetite. When these pieces were free to turn, one end pointed north. These might have been the first compasses. The compass was an important development for navigation and exploration, especially at sea. Before compasses, sailors had to depend on the Sun or the stars to know in which direction they were going.

Magnets

A piece of magnetite is a magnet. Magnets attract objects made of iron or steel, such as nails and paper clips. Magnets also can attract or repel other magnets. Every magnet has two ends, or poles. One end is called the north pole and the other is the south pole. As shown in **Figure 1,** a north magnetic pole always repels other north poles and always attracts south poles. Likewise, a south pole always repels other south poles and attracts north poles.

Two north poles repel

Two south poles repel

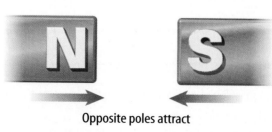

Figure 1 Two north poles or two south poles repel each other. North and south magnetic poles are attracted to each other.

Opposite poles attract

The Magnetic Field You have to handle a pair of magnets for only a short time before you can feel that magnets attract or repel without touching each other. How can a magnet cause an object to move without touching it? Recall that a force is a push or a pull that can cause an object to move. Just like gravitational and electric forces, a magnetic force can be exerted even when objects are not touching. And like these forces, the magnetic force becomes weaker as the magnets get farther apart.

This magnetic force is exerted through a **magnetic field.** Magnetic fields surround all magnets. If you sprinkle iron filings near a magnet, the iron filings will outline the magnetic field around the magnet. Take a look at **Figure 2.** The iron filings form a pattern of curved lines that start on one pole and end on the other. These curved lines are called magnetic field lines. Magnetic field lines help show the direction of the magnetic field.

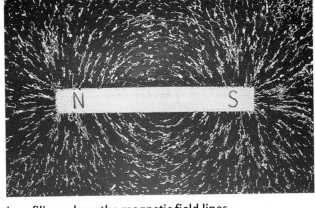

Iron filings show the magnetic field lines around a bar magnet.

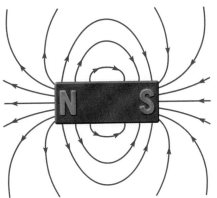

Magnetic field lines start at the north pole of the magnet and end on the south pole.

Reading Check *What is the evidence that a magnetic field exists?*

Magnetic field lines begin at a magnet's north pole and end on the south pole, as shown in **Figure 2.** The field lines are close together where the field is strong and get farther apart as the field gets weaker. As you can see in the figures, the magnetic field is strongest close to the magnetic poles and grows weaker farther from the poles.

Field lines that curve toward each other show attraction. Field lines that curve away from each other show repulsion. **Figure 3** illustrates the magnetic field lines between a north and a south pole and the field lines between two north poles.

Figure 2 A magnetic field surrounds a magnet. Where the magnetic field lines are close together, the field is strong.
Determine *for this magnet where the strongest field is.*

Attraction

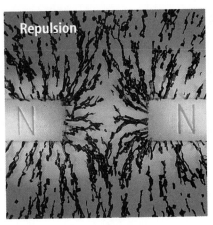

Repulsion

Figure 3 Magnetic field lines show attraction and repulsion.
Explain *what the field between two south poles would look like.*

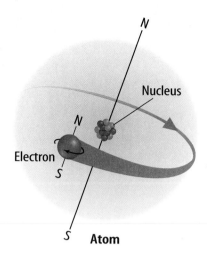

Figure 4 Movement of electrons produces magnetic fields. **Describe** *what two types of motion are shown in the illustration.*

Figure 5 Some materials can become temporary magnets.

Making Magnetic Fields Only certain materials, such as iron, can be made into magnets that are surrounded by a magnetic field. How are magnetic fields made? A moving electric charge, such as a moving electron, creates a magnetic field.

Inside every magnet are moving charges. All atoms contain negatively charged particles called electrons. Not only do these electrons swarm around the nucleus of an atom, they also spin, as shown in **Figure 4.** Because of its movement, each electron produces a magnetic field. The atoms that make up magnets have their electrons arranged so that each atom is like a small magnet. In a material such as iron, a large number of atoms will have their magnetic fields pointing in the same direction. This group of atoms, with their fields pointing in the same direction, is called a **magnetic domain.**

A material that can become magnetized, such as iron or steel, contains many magnetic domains. When the material is not magnetized, these domains are oriented in different directions, as shown in **Figure 5A.** The magnetic fields created by the domains cancel, so the material does not act like a magnet.

A magnet contains a large number of magnetic domains that are lined up and pointing in the same direction. Suppose a strong magnet is held close to a material such as iron or steel. The magnet causes the magnetic field in many magnetic domains to line up with the magnet's field, as shown in **Figure 5B.** As you can see in **Figure 5C** this process magnetizes paper clips.

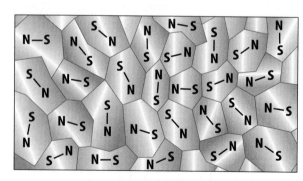

A Microscopic sections of iron and steel act as tiny magnets. Normally, these domains are oriented randomly and their magnetic fields cancel each other.

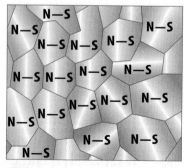

B When a strong magnet is brought near the material, the domains line up, and their magnetic fields add together.

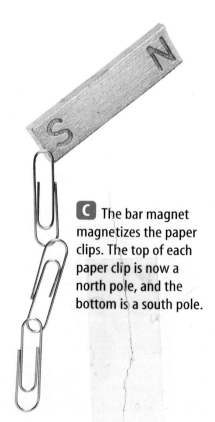

C The bar magnet magnetizes the paper clips. The top of each paper clip is now a north pole, and the bottom is a south pole.

Earth's Magnetic Field

Magnetism isn't limited to bar magnets. Earth has a magnetic field, as shown in **Figure 6.** The region of space affected by Earth's magnetic field is called the **magnetosphere** (mag NEE tuh sfihr). This deflects most of the charged particles from the Sun. The origin of Earth's magnetic field is thought to be deep within Earth in the outer core layer. One theory is that movement of molten iron in the outer core is responsible for generating Earth's magnetic field. The shape of Earth's magnetic field is similar to that of a huge bar magnet tilted about 11° from Earth's geographic north and south poles.

Figure 6 Earth has a magnetic field similar to the field of a bar magnet.

Applying Science

Finding the Magnetic Declination

The north pole of a compass points toward the magnetic pole, rather than true north. Imagine drawing a line between your location and the north pole, and a line between your location and the magnetic pole. The angle between these two lines is called the magnetic declination. Magnetic declination must be known if you need to know the direction to true north. However, the magnetic declination changes depending on your position.

Identifying the Problem

Suppose your location is at 50° N and 110° W. The location of the north pole is at 90° N and 110° W, and the location of the magnetic pole is at about 80° N and 105° W. What is the magnetic declination angle at your location?

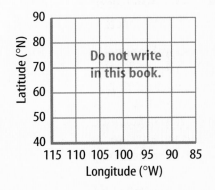

Latitude (°N) / Longitude (°W)

Do not write in this book.

Solving the Problem
1. Draw and label a graph like the one shown above.
2. On the graph, plot your location, the location of the magnetic pole, and the location of the north pole.
3. Draw a line from your location to the north pole, and a line from your location to the magnetic pole.
4. Using a protractor, measure the angle between the two lines.

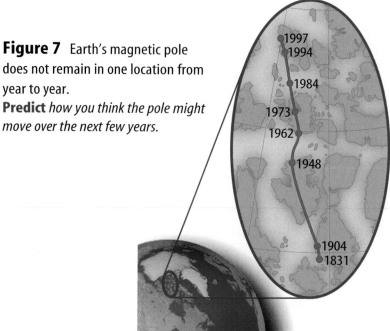

Figure 7 Earth's magnetic pole does not remain in one location from year to year.

Predict *how you think the pole might move over the next few years.*

1997
1994
1984
1973
1962
1948
1904
1831

Observing Magnetic Fields

Procedure

1. Place **iron filings** in a **plastic petri dish.** Cover the dish and seal it with **clear tape.**
2. Collect **several magnets.** Place the magnets on the table and hold the dish over each one. Draw a diagram of what happens to the filings in each case.
3. Arrange two or more magnets under the dish. Observe the pattern of the filings.

Analysis

1. What happens to the filings close to the poles? Far from the poles?
2. Compare the fields of the individual magnets. How can you tell which magnet is strongest? Weakest?

INTEGRATE Life Science

Nature's Magnets Honeybees, rainbow trout, and homing pigeons have something in common with sailors and hikers. They take advantage of magnetism to find their way. Instead of using compasses, these animals and others have tiny pieces of magnetite in their bodies. These pieces are so small that they may contain a single magnetic domain. Scientists have shown that some animals use these natural magnets to detect Earth's magnetic field. They appear to use Earth's magnetic field, along with other clues like the position of the Sun or stars, to help them navigate.

Earth's Changing Magnetic Field Earth's magnetic poles do not stay in one place. The magnetic pole in the north today, as shown in **Figure 7,** is in a different place from where it was 20 years ago. In fact, not only does the position of the magnetic poles move, but Earth's magnetic field sometimes reverses direction. For example, 700 thousand years ago, a compass needle that now points north would point south. During the past 20 million years, Earth's magnetic field has reversed direction more than 70 times. The magnetism of ancient rocks contains a record of these magnetic field changes. When some types of molten rock cool, magnetic domains of iron in the rock line up with Earth's magnetic field. After the rock cools, the orientation of these domains is frozen into position. Consequently, these old rocks preserve the orientation of Earth's magnetic field as it was long ago.

Figure 8 The compass needles align with the magnetic field lines around the magnet.
Explain *what happens to the compass needles when the bar magnet is removed.*

The Compass A compass needle is a small bar magnet with a north and south magnetic pole. In a magnetic field, a compass needle rotates until it is aligned with the magnetic field line at its location. **Figure 8** shows how the orientation of a compass needle depends on its location around a bar magnet.

Earth's magnetic field also causes a compass needle to rotate. The north pole of the compass needle points toward Earth's magnetic pole that is in the north. This magnetic pole is actually a magnetic south pole. Earth's magnetic field is like that of a bar magnet with the magnet's south pole near Earth's north pole.

Summary

Magnets

- A magnet has a north pole and a south pole.
- Like magnetic poles repel each other; unlike poles attract each other.
- A magnet is surrounded by a magnetic field that exerts forces on other magnets.
- Some materials are magnetic because their atoms behave like magnets.

Earth's Magnetic Field

- Earth is surrounded by a magnetic field similar to the field around a bar magnet.
- Earth's magnetic poles move slowly, and sometimes change places. Earth's magnetic poles now are close to Earth's geographic poles.

Self Check

1. **Explain** why atoms behave like magnets.
2. **Explain** why magnets attract iron but do not attract paper.
3. **Describe** how the behavior of electric charges is similar to that of magnetic poles.
4. **Determine** where the field around a magnet is the strongest and where it is the weakest.
5. **Think Critically** A horseshoe magnet is a bar magnet bent into the shape of the letter U. When would two horseshoe magnets attract each other? Repel? Have little effect?

Applying Skills

6. **Communicate** Ancient sailors navigated by using the Sun, stars, and following a coastline. Explain how the development of the compass would affect the ability of sailors to navigate.

Make a C✦mpass

A valuable tool for hikers and campers is a compass. Almost 1,000 years ago, Chinese inventors found a way to magnetize pieces of iron. They used this method to manufacture compasses. You can use the same procedure to make a compass.

◉ Real-World Question

How do you construct a compass?

Goals
- ■ **Observe** induced magnetism.
- ■ **Build** a compass.

Materials
petri dish	tape
clear bowl	marker
water	paper
sewing needle	plastic spoon
magnet	*Alternate material*

Safety Precautions 🥽 🧤 🤝

◉ Procedure

1. Reproduce the circular protractor shown. Tape it under the bottom of your dish so it can be seen but not get wet. Add water until the dish is half full.

2. Mark one end of the needle with a marker. Magnetize a needle by placing it on the magnet aligned north and south for 1 min.

3. Float the needle in the dish using a plastic spoon to lower the needle carefully onto the water. Turn the dish so the marked part of the needle is above the 0° mark. This is your compass.

4. Bring the magnet near your compass. Observe how the needle reacts. Measure the angle the needle turns.

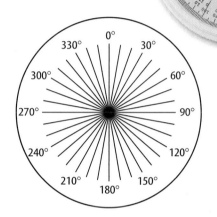

◉ Conclude and Apply

1. **Explain** why the marked end of the needle always pointed the same way in step 3, even though you rotated the dish.

2. **Describe** the behavior of the compass when the magnet was brought close.

3. **Observe** the marked end of your needle. Does it point to the north or south pole of the bar magnet? **Infer** whether the marked end of your needle is a north or a south pole. How do you know?

𝒞ommunicating
Your Data

Make a half-page insert that will go into a wilderness survival guide to describe the procedure for making a compass. Share your half-page insert with your classmates. **For more help, refer to the** Science Skill Handbook.

Electricity and Magnetism

Current Can Make a Magnet

Magnetic fields are produced by moving electric charges. Electrons moving around the nuclei of atoms produce magnetic fields. The motion of these electrons causes some materials, such as iron, to be magnetic. You cause electric charges to move when you flip a light switch or turn on a portable CD player. When electric current flows in a wire, electric charges move in the wire. As a result, a wire that contains an electric current also is surrounded by a magnetic field. **Figure 9A** shows the magnetic field produced around a wire that carries an electric current.

Electromagnets Look at the magnetic field lines around the coils of wire in **Figure 9B.** The magnetic fields around each coil of wire add together to form a stronger magnetic field inside the coil. When the coils are wrapped around an iron core, the magnetic field of the coils magnetizes the iron. The iron then becomes a magnet, which adds to the strength of the magnetic field inside the coil. A current-carrying wire wrapped around an iron core is called an **electromagnet,** as shown in **Figure 9C.**

Figure 9 A current-carrying wire produces a magnetic field.

as you read

What You'll Learn
- **Explain** how electricity can produce motion.
- **Explain** how motion can produce electricity.

Why It's Important
Electricity and magnetism enable electric motors and generators to operate.

🔍 **Review Vocabulary**
electric current: the flow of electric charge

New Vocabulary
- electromagnet
- motor
- aurora
- generator
- alternating current
- direct current
- transformer

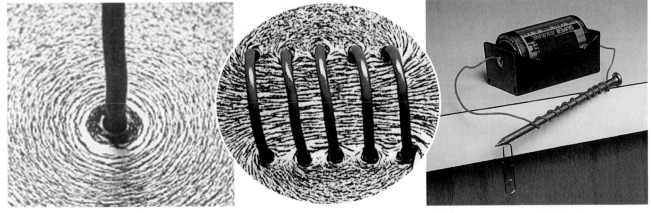

A Iron particles show the magnetic field lines around a current-carrying wire.

B When a wire is wrapped in a coil, the field inside the coil is made stronger.

C An iron core inside the coils increases the magnetic field because the core becomes magnetized.

Figure 10 An electric doorbell uses an electromagnet. Each time the electromagnet is turned on, the hammer strikes the bell. **Explain** *how the electromagnet is turned off.*

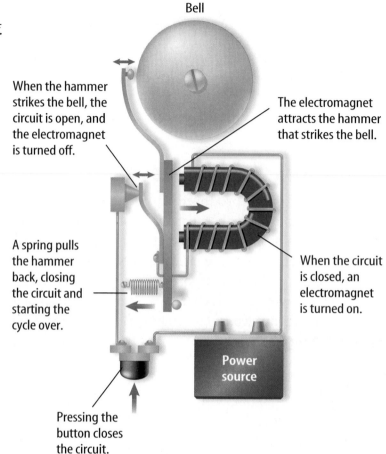

Bell

When the hammer strikes the bell, the circuit is open, and the electromagnet is turned off.

The electromagnet attracts the hammer that strikes the bell.

A spring pulls the hammer back, closing the circuit and starting the cycle over.

When the circuit is closed, an electromagnet is turned on.

Power source

Pressing the button closes the circuit.

Assembling an Electromagnet

Procedure

1. Wrap a **wire** around a **16-penny steel nail** ten times. Connect one end of the wire to a **D-cell battery,** as shown in **Figure 9C.** Leave the other end loose until you use the electromagnet. **WARNING:** *When current is flowing in the wire, it can become hot over time.*

2. Connect the wire. Observe how many **paper clips** you can pick up with the magnet.

3. Disconnect the wire and rewrap the nail with 20 coils. Connect the wire and observe how many paper clips you can pick up. Disconnect the wire again.

Analysis

1. How many paper clips did you pick up each time? Did more coils make the electromagnet stronger or weaker?

2. Graph the number of coils versus number of paper clips attracted. Predict how many paper clips would be picked up with five coils of wire. Check your prediction.

Try at Home

Using Electromagnets The magnetic field of an electromagnet is turned on or off when the electric current is turned on or off. By changing the current, the strength and direction of the magnetic field of an electromagnet can be changed. This has led to a number of practical uses for electromagnets. A doorbell, as shown in **Figure 10,** is a familiar use of an electromagnet. When you press the button by the door, you close a switch in a circuit that includes an electromagnet. The magnet attracts an iron bar attached to a hammer. The hammer strikes the bell. When the hammer strikes the bell, the hammer has moved far enough to open the circuit again. The electromagnet loses its magnetic field, and a spring pulls the iron bar and hammer back into place. This movement closes the circuit, and the cycle is repeated as long as the button is pushed.

Some gauges, such as the gas gauge in a car, use a galvanometer to move the gauge pointer. **Figure 11** shows how a galvanometer makes a pointer move. Ammeters and voltmeters used to measure current and voltage in electric circuits also use galvanometers, as shown in **Figure 11.**

Figure 11

The gas gauge in a car uses a device called a galvanometer to make the needle of the gauge move. Galvanometers are also used in other measuring devices. A voltmeter uses a galvanometer to measure the voltage in a electric circuit. An ammeter uses a galvanometer to measure electric current. Multimeters can be used as an ammeter or voltmeter by turning a switch.

A galvanometer has a pointer attached to a coil that can rotate between the poles of a permanent magnet. When a current flows through the coil, it becomes an electromagnet. Attraction and repulsion between the magnetic poles of the electromagnet and the poles of the permanent magnet makes the coil rotate. The amount of rotation depends on the amount of current in the coil.

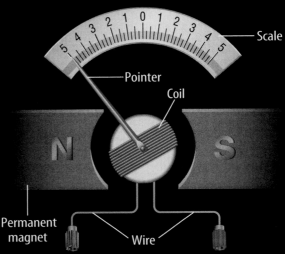

To measure the current in a circuit an ammeter is used. An ammeter contains a galvanometer and has low resistance. To measure current, an ammeter is connected in series in the circuit, so all the current in the circuit flows through it. The greater the current in the circuit, the more the needle moves.

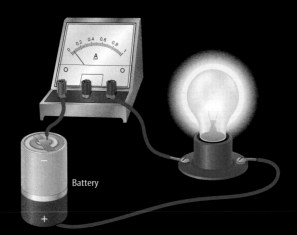

To measure the voltage in a circuit a voltmeter is used. A voltmeter also contains a galvanometer and has high resistance. To measure voltage, a voltmeter is connected in parallel in the circuit, so almost no current flows through it. The higher the voltage in the circuit, the more the needle moves.

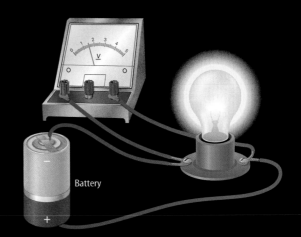

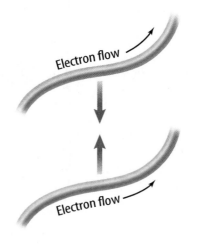
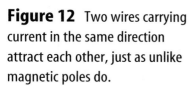
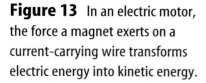

Magnets Push and Pull Currents

Look around for electric appliances that produce motion, such as a fan. How does the electric energy entering the fan become transformed into the kinetic energy of the moving fan blades? Recall that current-carrying wires produce a magnetic field. This magnetic field behaves the same way as the magnetic field that a magnet produces. Two current-carrying wires can attract each other as if they were two magnets, as shown in **Figure 12.**

Figure 12 Two wires carrying current in the same direction attract each other, just as unlike magnetic poles do.

Electric Motor Just as two magnets exert a force on each other, a magnet and a current-carrying wire exert forces on each other. The magnetic field around a current-carrying wire will cause it to be pushed or pulled by a magnet, depending on the direction the current is flowing in the wire. As a result, some of the electric energy carried by the current is converted into kinetic energy of the moving wire, as shown on the left in **Figure 13.** Any device that converts electric energy into kinetic energy is a **motor.** To keep a motor running, the current-carrying wire is formed into a loop so the magnetic field can force the wire to spin continually, as shown on the right in **Figure 13.**

Figure 13 In an electric motor, the force a magnet exerts on a current-carrying wire transforms electric energy into kinetic energy.

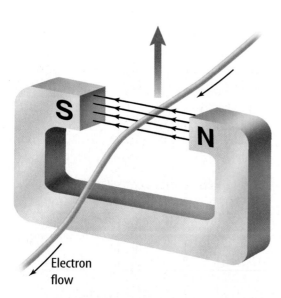

A magnetic field like the one shown will push a current-carrying wire upward.

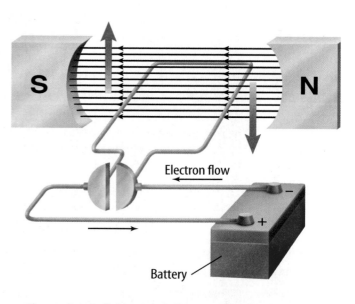

The magnetic field exerts a force on the wire loop, causing it to spin as long as current flows in the loop.

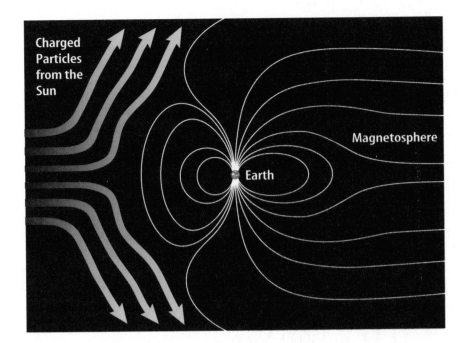

Figure 14 Earth's magneto-sphere deflects most of the charged particles streaming from the Sun. **Explain** why the magnetosphere is stretched away from the Sun.

Earth's Magnetosphere The Sun emits charged particles that stream through the solar system like an enormous electric current. Just like a current-carrying wire is pushed or pulled by a magnetic field, Earth's magnetic field pushes and pulls on the electric current generated by the Sun. This causes most of the charged particles in this current to be deflected so they never strike Earth, as shown in **Figure 14.** As a result, living things on Earth are protected from damage that might be caused by these charged particles. At the same time, the solar current pushes on Earth's magnetosphere so it is stretched away from the Sun.

The Aurora Sometimes the Sun ejects a large number of charged parti-cles all at once. Most of these charged particles are deflected by Earth's mag-netosphere. However, some of the ejected particles from the Sun produce other charged particles in Earth's outer atmosphere. These charged particles spiral along Earth's magnetic field lines toward Earth's magnetic poles. There they collide with atoms in the atmos-phere. These collisions cause the atoms to emit light. The light emitted causes a display known as the **aurora** (uh ROR uh), as shown in **Figure 15.** In north-ern latitudes, the aurora sometimes is called the northern lights.

Figure 15 An aurora is a natural light show that occurs far north and far south.

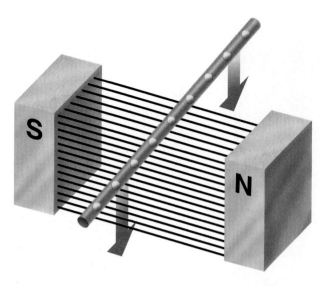

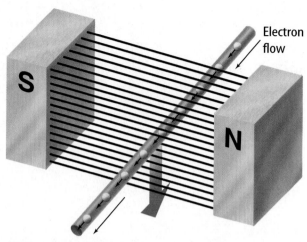

Electron flow

If a wire is pulled through a magnetic field, the electrons in the wire also move downward.

The magnetic field then exerts a force on the moving electrons, causing them to move along the wire.

Figure 16 When a wire is made to move through a magnetic field, an electric current can be produced in the wire.

Figure 17 In a generator, a power source spins a wire loop in a magnetic field. Every half turn, the current will reverse direction. This type of generator supplies alternating current to the lightbulb.

Using Magnets to Create Current

In an electric motor, a magnetic field turns electricity into motion. A device called a **generator** uses a magnetic field to turn motion into electricity. Electric motors and electric generators both involve conversions between electric energy and kinetic energy. In a motor, electric energy is changed into kinetic energy. In a generator, kinetic energy is changed into electric energy. **Figure 16** shows how a current can be produced in a wire that moves in a magnetic field. As the wire moves, the electrons in the wire also move in the same direction, as shown on the left. The magnetic field exerts a force on the moving electrons that pushes them along the wire on the right, creating an electric current.

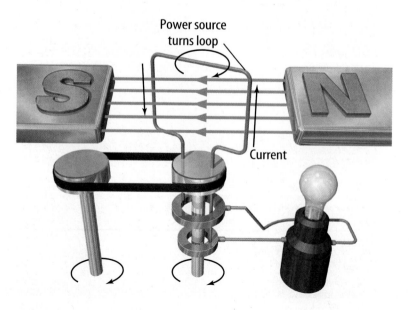

Power source turns loop

Current

Electric Generators To produce electric current, the wire is fashioned into a loop, as in **Figure 17.** A power source provides the kinetic energy to spin the wire loop. With each half turn, the current in the loop changes direction. This causes the current to alternate from positive to negative. Such a current is called an **alternating current** (AC). In the United States, electric currents change from positive to negative to positive 60 times each second.

Types of Current A battery produces direct current instead of alternating current. In a **direct current** (DC) electrons flow in one direction. In an alternating current, electrons change their direction of movement many times each second. Some generators are built to produce direct current instead of alternating current.

Reading Check *What type of currents can be produced by a generator?*

Power Plants Electric generators produce almost all of the electric energy used all over the world. Small generators can produce energy for one household, and large generators in electric power plants can provide electric energy for thousands of homes. Different energy sources such as gas, coal, and water are used to provide the kinetic energy to rotate coils of wire in a magnetic field. Coal-burning power plants, like the one pictured in **Figure 18,** are the most common. More than half of the electric energy generated by power plants in the United States comes from burning coal.

Voltage The electric energy produced at a power plant is carried to your home in wires. Recall that voltage is a measure of how much energy the electric charges in a current are carrying. The electric transmission lines from electric power plants transmit electric energy at a high voltage of about 700,000 V. Transmitting electric energy at a low voltage is less efficient because more electric energy is converted into heat in the wires. However, high voltage is not safe for use in homes and businesses. A device is needed to reduce the voltage.

Science Online

Topic: Power Plants
Visit bookn.msscience.com for Web links to more information about the different types of power plants used in your region of the country.

Activity Describe the different types of power plants.

Figure 18 Coal-burning power plants supply much of the electric energy for the world.

Figure 19 Electricity travels from a generator to your home.

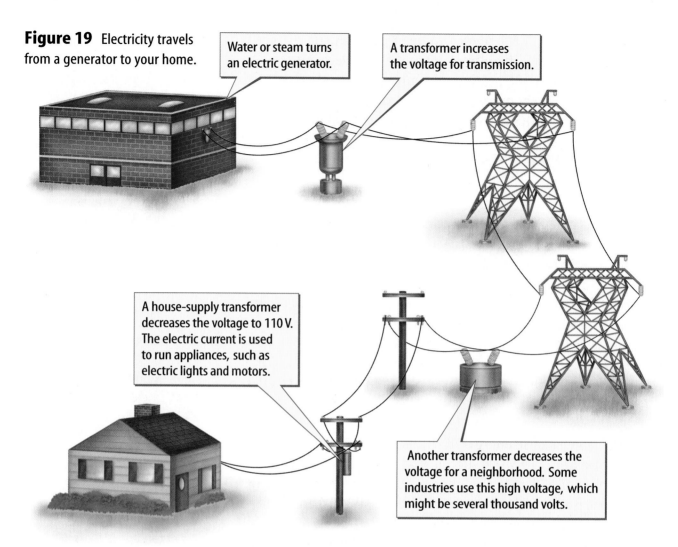

Water or steam turns an electric generator.

A transformer increases the voltage for transmission.

A house-supply transformer decreases the voltage to 110 V. The electric current is used to run appliances, such as electric lights and motors.

Another transformer decreases the voltage for a neighborhood. Some industries use this high voltage, which might be several thousand volts.

Changing Voltage

A **transformer** is a device that changes the voltage of an alternating current with little loss of energy. Transformers are used to increase the voltage before transmitting an electric current through the power lines. Other transformers are used to decrease the voltage to the level needed for home or industrial use. Such a power system is shown in **Figure 19.** Transformers also are used in power adaptors. For battery-operated devices, a power adaptor must change the 120 V from the wall outlet to the same voltage produced by the device's batteries.

Reading Check *What does a transformer do?*

A transformer usually has two coils of wire wrapped around an iron core, as shown in **Figure 20.** One coil is connected to an alternating current source. The current creates a magnetic field in the iron core, just like in an electromagnet. Because the current is alternating, the magnetic field it produces also switches direction. This alternating magnetic field in the core then causes an alternating current in the other wire coil.

Figure 20 A transformer can increase or decrease voltage. The ratio of input coils to output coils equals the ratio of input voltage to output voltage.
Determine *the output voltage if the input voltage is 60 V.*

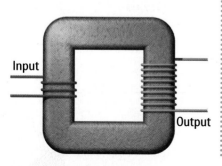

Input

Output

The Transformer Ratio Whether a transformer increases or decreases the input voltage depends on the number of coils on each side of the transformer. The ratio of the number of coils on the input side to the number of coils on the output side is the same as the ratio of the input voltage to the output voltage. For the transformer in **Figure 20,** the ratio of the number of coils on the input side to the number of coils on the output side is three to nine, or one to three. If the input voltage is 60 V, the output voltage will be 180 V.

In a transformer the voltage is greater on the side with more coils. If the number of coils on the input side is greater than the number of coils on the output side, the voltage is decreased. If the number of coils on the input side is less than the number on the output side, the voltage is increased.

Superconductors

Electric current can flow easily through materials, such as metals, that are electrical conductors. However, even in conductors, there is some resistance to this flow and heat is produced as electrons collide with atoms in the material.

Unlike an electrical conductor, a material known as a superconductor has no resistance to the flow of electrons. Superconductors are formed when certain materials are cooled to low temperatures. For example, aluminum becomes a superconductor at about −272ºC. When an electric current flows through a superconductor, no heat is produced and no electric energy is converted into heat.

Superconductors and Magnets Superconductors also have other unusual properties. For example, a magnet is repelled by a superconductor. As the magnet gets close to the superconductor, the superconductor creates a magnetic field that is opposite to the field of the magnet. The field created by the superconductor can cause the magnet to float above it, as shown in **Figure 21.**

INTEGRATE History

The Currents War In the late 1800s, electric power was being transmitted using a direct-current transmission system developed by Thomas Edison. To preserve his monopoly, Edison launched a public-relations war against the use of alternating-current power transmission, developed by George Westinghouse and Nikola Tesla. However, by 1893, alternating current transmission had been shown to be more efficient and economical, and quickly became the standard.

Figure 21 A small magnet floats above a superconductor. The magnet causes the superconductor to produce a magnetic field that repels the magnet.

Figure 22 The particle accelerator at Fermi National Accelerator Laboratory near Batavia, Illinois, accelerates atomic particles to nearly the speed of light. The particles travel in a beam only a few millimeters in diameter. Magnets made of superconductors keep the beam moving in a circular path about 2 km in diameter.

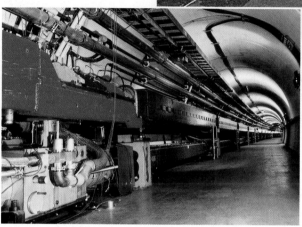

Figure 23 A patient is being placed inside an MRI machine. The strong magnetic field inside the machine enables images of tissues inside the patient's body to be made.

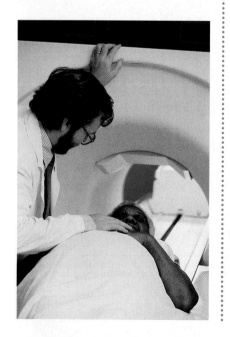

Using Superconductors Large electric currents can flow through electromagnets made from superconducting wire and can produce extremely strong magnetic fields. The particle accelerator shown in **Figure 22** uses more than 1,000 superconducting electromagnets to help accelerate subatomic particles to nearly the speed of light.

Other uses for superconductors are being developed. Transmission lines made from a superconductor could transmit electric power over long distances without having any electric energy converted to heat. It also may be possible to construct extremely fast computers using microchips made from superconductor materials.

Magnetic Resonance Imaging

INTEGRATE Health   A method called magnetic resonance imaging, or MRI, uses magnetic fields to create images of the inside of a human body. MRI images can show if tissue is damaged or diseased, and can detect the presence of tumors.

Unlike X-ray imaging, which uses X-ray radiation that can damage tissue, MRI uses a strong magnetic field and radio waves. The patient is placed inside a machine like the one shown in **Figure 23.** Inside the machine an electromagnet made from superconductor materials produces a magnetic field more than 20,000 times stronger than Earth's magnetic field.

Producing MRI Images About 63 percent of all the atoms in your body are hydrogen atoms. The nucleus of a hydrogen atom is a proton, which behaves like a tiny magnet. The strong magnetic field inside the MRI tube causes these protons to line up along the direction of the field. Radio waves are then applied to the part of the body being examined. The protons absorb some of the energy in the radio waves, and change the direction of their alignment.

When the radio waves are turned off, the protons realign themselves with the magnetic field and emit the energy they absorbed. The amount of energy emitted depends on the type of tissue in the body. This energy emitted is detected and a computer uses this information to form an image, like the one shown in **Figure 24.**

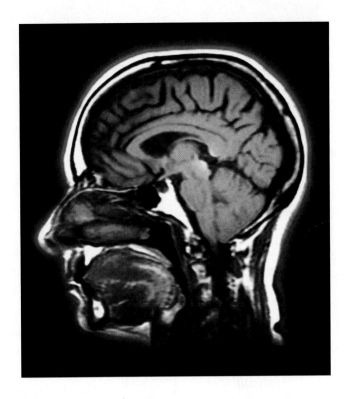

Figure 24 This MRI image shows a side view of the brain.

Connecting Electricity and Magnetism Electric charges and magnets are related to each other. Moving electric charges produce magnetic fields, and magnetic fields exert forces on moving electric charges. It is this connection that enables electric motors and generators to operate.

section 2 review

Summary

Electromagnets

- A current-carrying wire is surrounded by a magnetic field.
- An electromagnet is made by wrapping a current-carrying wire around an iron core.

Motors, Generators, and Transformers

- An electric motor transforms electrical energy into kinetic energy. An electric motor rotates when current flows in a wire loop that is surrounded by a magnetic field.
- An electric generator transforms kinetic energy into electrical energy. A generator produces current when a wire loop is rotated in a magnetic field.
- A transformer changes the voltage of an alternating current.

Self-Check

1. **Describe** how the magnetic field of an electromagnet depends on the current and the number of coils.

2. **Explain** how a transformer works.

3. **Describe** how a magnetic field affects a current-carrying wire.

4. **Describe** how alternating current is produced.

5. **Think Critically** What are some advantages and disadvantages to using superconductors as electric transmission lines?

Applying Math

6. **Calculate Ratios** A transformer has ten turns of wire on the input side and 50 turns of wire on the output side. If the input voltage is 120 V, what will the output voltage be?

How does an electric m🌸tor work?

Goals

- **Assemble** a small electric motor.
- **Observe** how the motor works.

Materials

22-gauge enameled wire (4 m)
steel knitting needle
*steel rod
nails (4)
hammer
ceramic magnets (2)
18-gauge insulated wire (60 cm)
masking tape
fine sandpaper
approximately 15-cm square wooden board
wooden blocks (2)
6-V battery
*1.5-V batteries connected in a series (4)
wire cutters
*scissors
*Alternate materials

Safety Precautions

WARNING: *Hold only the insulated part of a wire when it is attached to the battery. Use care when hammering nails. After cutting the wire, the ends will be sharp.*

▶ Real-World Question

Electric motors are used in many appliances. For example, a computer contains a cooling fan and motors to spin the hard drive. A CD player contains electric motors to spin the CD. Some cars contain electric motors that move windows up and down, change the position of the seats, and blow warm or cold air into the car's interior. All these electric motors consist of an electromagnet and a permanent magnet. In this activity you will build a simple electric motor that will work for you. How can you change electric energy into motion?

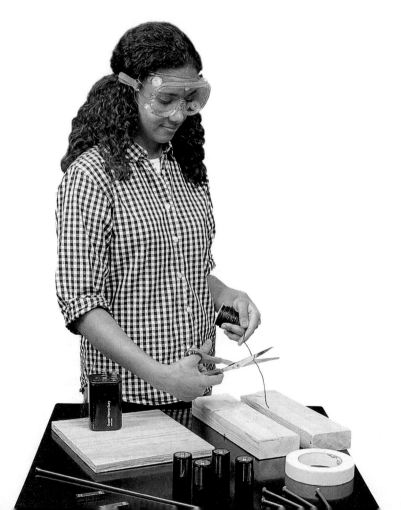

Procedure

1. Use sandpaper to strip the enamel from about 4 cm of each end of the 22-gauge wire.

2. Leaving the stripped ends free, make this wire into a tight coil of at least 30 turns. A D-cell battery or a film canister will help in forming the coil. Tape the coil so it doesn't unravel.

3. Insert the knitting needle through the coil. Center the coil on the needle. Pull the wire's two ends to one end of the needle.

4. Near the ends of the wire, wrap masking tape around the needle to act as insulation. Then tape one bare wire to each side of the needle at the spot where the masking tape is.

5. Tape a ceramic magnet to each block so that a north pole extends from one and a south pole from the other.

6. Make the motor. Tap the nails into the wood block as shown in the figure. Try to cross the nails at the same height as the magnets so the coil will be suspended between them.

7. Place the needle on the nails. Use bits of wood or folded paper to adjust the positions of the magnets until the coil is directly between the magnets. The magnets should be as close to the coil as possible without touching it.

8. Cut two 30-cm lengths of 18-gauge wire. Use sandpaper to strip the ends of both wires. Attach one wire to each terminal of the battery. Holding only the insulated part of each wire, place one wire against each of the bare wires taped to the needle to close the circuit. Observe what happens.

Conclude and Apply

1. **Describe** what happens when you close the circuit by connecting the wires. Were the results expected?

2. **Describe** what happens when you open the circuit.

3. **Predict** what would happen if you used twice as many coils of wire.

Communicating Your Data

Compare your conclusions with other students in your class. **For more help, refer to the** Science Skill Handbook.

"Aagjuuk[1] and Sivulliit[2]"
from Intellectual Culture of the Copper Eskimos
by Knud Rasmussen, told by Tatilgak

The following are "magic words" that are spoken before the Inuit (IH noo wut) people go seal hunting. Inuit are native people that live in the arctic region. Because the Inuit live in relative darkness for much of the winter, they have learned to find their way by looking at the stars to guide them. The poem is about two constellations that are important to the Inuit people because their appearance marks the end of winter when the Sun begins to appear in the sky again.

By which way, I wonder the mornings—
You dear morning, get up!
See I am up!
By which way, I wonder,
the constellation *Aagjuuk* rises up in the sky?
By this way—perhaps—by the morning
It rises up!

Morning, you dear morning, get up!
See I am up!
By which way, I wonder,
the constellation *Sivulliit*
Has risen to the sky?
By this way—perhaps—by the morning.
It rises up!

[1]Inuit name for the constellation of stars called Aquila (A kwuh luh)
[2]Inuit name for the constellation of stars called Bootes (boh OH teez)

Understanding Literature

Ethnography Ethnography is a description of a culture. To write an ethnography, an ethnographer collects cultural stories, poems, or other oral tales from the culture that he or she is studying. Why must the Inuit be skilled in navigation?

Respond to the Reading

1. How can you tell the importance of constellations to the Inuit for telling direction?
2. How is it possible that the Inuit could see the constellations in the morning sky?
3. **Linking Science and Writing** Research the constellations in the summer sky in North America and write a paragraph describing the constellations that would help you navigate from south to north.

INTEGRATE Physics Earth's magnetic field causes the north pole of a compass needle to point in a northerly direction. Using a compass helps a person to navigate and find his or her way. However, at the far northern latitudes where the Inuit live, a compass becomes more difficult to use. Some Inuit live north of Earth's northern magnetic pole. In these locations a compass needle points in a southerly direction. As a result, the Inuit developed other ways to navigate.

Reviewing Main Ideas

Section 1 What is magnetism?

1. All magnets have two poles—north and south. Like poles repel each other and unlike poles attract.

2. A magnet is surrounded by a magnetic field that exerts forces on other magnets.

3. Atoms in magnetic materials are magnets. These materials contain magnetic domains which are groups of atoms whose magnetic poles are aligned.

4. Earth is surrounded by a magnetic field similar to the field around a bar magnet.

Section 2 Electricity and Magnetism

1. Electric current creates a magnetic field. Electromagnets are made from a coil of wire that carries a current, wrapped around an iron core.

2. A magnetic field exerts a force on a moving charge or a current-carrying wire.

3. Motors transform electric energy into kinetic energy. Generators transform kinetic energy into electric energy.

4. Transformers are used to increase and decrease voltage in AC circuits.

Visualizing Main Ideas

Copy and complete the following concept map on magnets.

Magnets

are made from — Magnetic materials — in which — Moving electrons in atoms — produce — ○ — that line up to make — ○

are used by — Electric motors — in which — ○ — generates — ○ — that produces — Kinetic energy

are used by — Generators — in which — ○ — causes — A wire loop to rotate — that generates — ○

Using Vocabulary

alternating current p. 50	magnetic domain p. 40
aurora p. 49	magnetic field p. 39
direct current p. 51	magnetosphere p. 41
electromagnet p. 45	motor p. 48
generator p. 50	transformer p. 52

Explain the relationship that exists between each set of vocabulary words below.

1. generator—transformer

2. magnetic force—magnetic field

3. alternating current—direct current

4. current—electromagnet

5. motor—generator

6. electron—magnetism

7. magnetosphere—aurora

8. magnet—magnetic domain

Checking Concepts

Choose the word or phrase that best answers the question.

9. What can iron filings be used to show?
   A) magnetic field   **C)** gravitational field
   B) electric field   **D)** none of these

10. Why does the needle of a compass point to magnetic north?
    A) Earth's north pole is strongest.
    B) Earth's north pole is closest.
    C) Only the north pole attracts compasses.
    D) The compass needle aligns itself with Earth's magnetic field.

11. What will the north poles of two bar magnets do when brought together?
    A) attract
    B) create an electric current
    C) repel
    D) not interact

12. How many poles do all magnets have?
    A) one   **C)** three
    B) two   **D)** one or two

13. When a current-carrying wire is wrapped around an iron core, what can it create?
    A) an aurora   **C)** a generator
    B) a magnet   **D)** a motor

14. What does a transformer between utility wires and your house do?
    A) increases voltage
    B) decreases voltage
    C) leaves voltage the same
    D) changes DC to AC

Use the figure below to answer question 15.

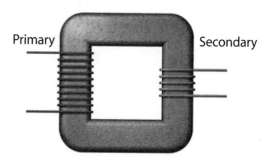

15. For this transformer which of the following describes how the output voltage compares with the input voltage?
    A) larger   **C)** smaller
    B) the same   **D)** zero voltage

16. Which energy transformation occurs in an electric motor?
    A) electrical to kinetic
    B) electrical to thermal
    C) potential to kinetic
    D) kinetic to electrical

17. What prevents most charged particles from the Sun from hitting Earth?
    A) the aurora
    B) Earth's magnetic field
    C) high-altitude electric fields
    D) Earth's atmosphere

Thinking Critically

18. **Concept Map** Explain how a doorbell uses an electromagnet by placing the following phrases in the cycle concept map: *circuit open, circuit closed, electromagnet turned on, electromagnet turned off, hammer attracted to magnet and strikes bell,* and *hammer pulled back by a spring.*

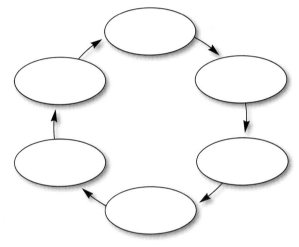

19. **Infer** A nail is magnetized by holding the south pole of a magnet against the head of the nail. Does the point of the nail become a north pole or a south pole? Include a diagram with your explanation.

20. **Explain** why an ordinary bar magnet doesn't rotate and align itself with Earth's magnetic field when you place it on a table.

21. **Determine** Suppose you were given two bar magnets. One magnet has the north and south poles labeled, and on the other magnet the magnetic poles are not labeled. Describe how you could use the labeled magnet to identify the poles of the unlabeled magnet.

22. **Explain** A bar magnet touches a paper clip that contains iron. Explain why the paper clip becomes a magnet that can attract other paper clips.

23. **Explain** why the magnetic field produced by an electromagnet becomes stronger when the wire coils are wrapped around an iron core.

24. **Predict** Magnet A has a magnetic field that is three times as strong as the field around magnet B. If magnet A repels magnet B with a force of 10 N, what is the force that magnet B exerts on magnet A?

25. **Predict** Two wires carrying electric current in the same direction are side by side and are attracted to each other. Predict how the force between the wires changes if the current in both wires changes direction.

Performance Activities

26. **Multimedia Presentation** Prepare a multimedia presentation to inform your classmates on the possible uses of superconductors.

Applying Math

Use the table below to answer questions 27 and 28.

Transformer Properties

Transformer	Number of Input Coils	Number of Output Coils
R	4	12
S	10	2
T	3	6
U	5	10

27. **Input and Output Coils** According to this table, what is the ratio of the number of input coils to the number of output coils on transformer T?

28. **Input and Output Voltage** If the input voltage is 60 V, which transformer gives an output voltage of 12 V?

Part 1 | Multiple Choice

Record your answers on the answer sheet provided by your teacher or on a sheet of paper.

Use the figure below to answer questions 1 and 2.

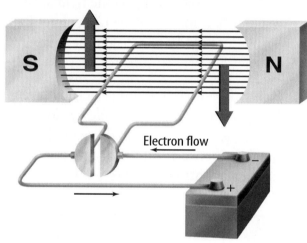

Electron flow

1. What is the device shown?
   A. electromagnet   C. electric motor
   B. generator   D. transformer

2. Which of the following best describes the function of this device?
   A. It transforms electrical energy into kinetic energy.
   B. It transforms kinetic energy into electrical energy.
   C. It increases voltage.
   D. It produces an alternating current.

3. How is an electromagnet different from a permanent magnet?
   A. It has north and south poles.
   B. It attracts magnetic substances.
   C. Its magnetic field can be turned off.
   D. Its poles cannot be reversed.

Test-Taking Tip

Check the Question Number For each question, double check that you are filling in the correct answer bubble for the question number you are working on.

4. Which of the following produces alternating current?
   A. electromagnet   C. generator
   B. superconductor   D. motor

5. Which statement about the domains in a magnetized substance is true?
   A. Their poles are in random directions.
   B. Their poles cancel each other.
   C. Their poles point in one direction.
   D. Their orientation cannot change.

Use the figure below to answer questions 6–8.

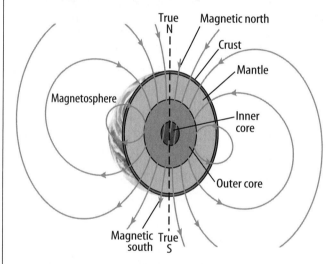

6. What is the region of space affected by Earth's magnetic field called?
   A. declination   C. aurora
   B. magnetosphere   D. outer core

7. What is the shape of Earth's magnetic field similar to?
   A. that of a horseshoe magnet
   B. that of a bar magnet
   C. that of a disk magnet
   D. that of a superconductor

8. In which of Earth's layers is Earth's magnetic field generated?
   A. crust   C. outer core
   B. mantle   D. inner core

Part 2 | Short Response/Grid In

Record your answers on the answer sheet provided by your teacher or on a sheet of paper.

Use the figure below to answer questions 9 and 10.

9. Explain why the compass needles are pointed in different directions.

10. What will happen to the compass needles when the bar magnet is removed? Explain why this happens.

11. Describe the interaction between a compass needle and a wire in which an electric current is flowing.

12. What are two ways to make the magnetic field of an electromagnet stronger?

13. The input voltage in a transformer is 100 V and the output voltage is 50 V. Find the ratio of the number of wire turns on the input coil to the number of turns on the output coil.

14. Explain how you could magnetize a steel screwdriver.

15. Suppose you break a bar magnet in two. How many magnetic poles does each of the pieces have?

16. Alnico is a mixture of steel, aluminum, nickel, and cobalt. It is very hard to magnetize. However, once magnetized, it remains magnetic for a long time. Explain why it would not be a good choice for the core of an electromagnet.

Part 3 | Open Ended

Record your answers on a sheet of paper.

17. Explain why the aurora occurs only near Earth's north and south poles.

18. Why does a magnet attract an iron nail to either of its poles, but attracts another magnet to only one of its poles?

19. A battery is connected to the input coil of a step-up transformer. Describe what happens when a lightbulb is connected to the output coil of the transformer.

20. Explain how electric forces and magnetic forces are similar.

Use the figure below to answer questions 21 and 22.

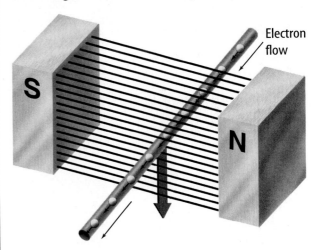

21. Describe the force that is causing the electrons to flow in the wire.

22. Infer how electrons would flow in the wire if the wire were pulled upward.

23. Explain why a nail containing iron can be magnetized, but a copper penny that contains no iron cannot be magnetized.

24. Every magnet has a north pole and a south pole. Where would the poles of a magnet that is in the shape of a disc be located?

The BIG Idea

Information is stored electronically as binary numbers in a computer.

SECTION 1
Electronics
Main Idea Sampling converts an analog signal into a series of numbers.

SECTION 2
Computers
Main Idea A computer carries out instructions contained in programs stored in the computer's memory.

Electronics and Computers

Deep in Thought?

You are looking at a brain—the brain of a computer. This is a microprocessor—a device that controls a computer. Even though this microprocessor is only a few centimeters on a side, it contains over a million microscopic circuits that enable it to store and process information very quickly.

Science Journal Describe three activities that you do using a computer.

Start-Up Activities

Electronic and Human Calculators

Imagine how your life would be different if you had been born before the invention of electronic devices. You could not watch television or use a computer. Besides providing entertainment, electronic devices and computers can make many tasks easier. For example, how much quicker is an electronic calculator than a human calculator?

1. Use a stopwatch to time how long it takes a volunteer to add the numbers 423, 21, 84, and 1,098.

2. Time how long it takes another volunteer to add these numbers using a calculator.

3. Repeat steps 1 and 2 this time asking the competitors to multiply 149 and 876.

4. Divide the time needed by the student calculator by the time needed by the calculator to solve each problem. How many times faster is the calculator?

5. **Think Critically** Write a paragraph describing which step in each calculation takes the most time.

Science Online | Preview this chapter's content and activities at bookn.msscience.com

---

FOLDABLES™
Study Organizer

Electronics and Computers
Make the following Foldable to help you identify what you already know and what you want to learn about electronics and computers.

STEP 1 **Fold** a vertical sheet of paper from side to side. Make the front edge about 1 cm shorter than the back edge.

STEP 2 **Turn** lengthwise and **fold** into thirds.

STEP 3 **Unfold and cut** only the top layer along both folds to make three tabs.

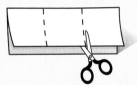

STEP 4 **Label** the tabs as shown.

Know Want Learned

Identify Questions Before you read the chapter, write what you know under the left tab and what you want to know under the middle tab. As you read the chapter, add to and correct what you have written. After you read the chapter, write what you have learned under the right tab of your Foldable.

Get Ready to Read

Make Connections

① **Learn It!** Make connections between what you read and what you already know. Connections can be based on personal experiences (text-to-self), what you have read before (text-to-text), or events in other places (text-to-world).

As you read, ask connecting questions. Are you reminded of a personal experience? Have you read about the topic before? Did you think of a person, a place, or an event in another part of the world?

② **Practice It!** Read the excerpt below and make connections to your own knowledge and experience.

Text-to-self:
What other electronic devices have you used that produce sounds or images?

Text-to-text:
What have read about electric current in other chapters?

Text-to-world:
Where are some places that loudspeakers are used?

> You have used another analog device if you have ever made a recording on a magnetic tape recorder. When voices or music are recorded on magnetic tape, the tape stores an analog signal of the sounds. When you play the tape, the tape recorder converts the analog signal to an electric current. This current changes smoothly with time, and causes a loudspeaker to vibrate, recreating the sounds for you to hear.
>
> — *from page 67*

③ **Apply It!** As you read this chapter, choose five words or phrases that make a connection to something you already know.

Reading Tip

Make connections with memorable events, places, or people in your life. The better the connection, the more you will remember.

Target Your Reading

Use this to focus on the main ideas as you read the chapter.

1 **Before you read** the chapter, respond to the statements below on your worksheet or on a numbered sheet of paper.
- Write an **A** if you **agree** with the statement.
- Write a **D** if you **disagree** with the statement.

2 **After you read** the chapter, look back to this page to see if you've changed your mind about any of the statements.
- If any of your answers changed, explain why.
- Change any false statements into true statements.
- Use your revised statements as a study guide.

Sciencenline

Print out a worksheet of this page at bookn.msscience.com

Before You Read A or D		Statement	After You Read A or D
	1	Any electric current can carry information.	
	2	A digital signal changes smoothly with time.	
	3	A digital signal can be represented by a series of numbers.	
	4	Modern integrated circuits might contain millions of vacuum tubes.	
	5	Information is stored on a computer as numbers that contain only the digits 0 and 1.	
	6	Computer programs are tiny electronic circuits that store information.	
	7	Everything a computer does is controlled by computer programs.	
	8	A hard disk uses magnetism to store information.	
	9	A CD uses magnetism to store information.	

Electronics

What You'll Learn

- **Compare and contrast** analog and digital signals.
- **Explain** how semiconductors are used in electronic devices.

Why It's Important

You use electronic devices every day to make your life easier and more enjoyable.

🔎 Review Vocabulary

crystal: a solid substance which has a regularly repeating internal arrangement of atoms

New Vocabulary

- electronic signal
- analog signal
- digital signal
- semiconductor
- diode
- transistor
- integrated circuit

Electronic Signals

You've popped some popcorn, put a video in the VCR, and turned off the lights. Now you're ready to watch a movie. The VCR, television, and lamp shown in **Figure 1** use electricity to operate. However, unlike the lamp, the VCR and the TV are electronic devices. An electronic device uses electricity to store, process, and transfer information.

The VCR and the TV use information recorded on the videotape to produce the images and sounds you see as a movie. As the videotape moves inside the VCR, it produces a changing electric current. This changing electric current is the information the VCR uses to send signals to the TV. The TV then uses these signals to produce the images you see and the sounds you hear.

A changing electric current that carries information is an **electronic signal.** The information can be used to produce sounds, images, printed words, numbers, or other data. For example, a changing electric current causes a loudspeaker to produce sound. If the electric current didn't change, no sound would be produced by the loudspeaker. There are two types of electronic signals—analog and digital.

Analog Signals Most TVs, VCRs, radios, and telephones process and transmit information that is in the form of analog electronic signals. An **analog signal** is a signal that varies smoothly in time. In an analog electronic signal the electric current increases or decreases smoothly in time, just as your hand can move smoothly up and down.

Electronic signals are not the only types of analog signals. An analog signal can be produced by something that varies in a smooth, continuous way and contains information. For example, a person's temperature changes smoothly and contains information about a person's health.

Figure 1 A VCR sends electronic signals to the TV, which uses the information in these signals to produce images and sound.

Figure 2 Clocks can be analog or digital devices.

The information displayed on an analog device such as this clock changes continuously.

On this digital clock, the displayed time jumps from one number to another.

Analog Devices The clock with hands shown in **Figure 2** is an example of an analog device. The hands move smoothly from one number to the next to represent the time of day. Fluid-filled and dial thermometers also are analog devices. In a fluid-filled thermometer, the height of the fluid column smoothly rises or falls as the temperature changes. In a dial thermometer, a spring smoothly expands or contracts as the temperature changes.

You have used another analog device if you ever have made a recording on a magnetic tape recorder. When voices or music are recorded on magnetic tape, the tape stores an analog signal of the sounds. When you play the tape, the tape recorder converts the analog signal to an electric current. This current changes smoothly with time and causes a loudspeaker to vibrate, recreating the sounds for you to hear.

Digital Signals Some devices, such as CD players, use a different kind of electronic signal called a digital signal. Unlike an analog signal, a **digital signal** does not vary smoothly, but changes in jumps or steps. If each jump is represented by a number, a digital signal can be represented by a series of numbers.

Reading Check *How is a digital signal different from an analog signal?*

You might have a digital clock or watch similar to the one shown on the right in **Figure 2** that displays the time as numbers. The display changes from 6:29 to 6:30 in a single jump, rather than sweeping smoothly from second to second. You might have seen digital thermometers that display temperature as a number. Some digital thermometers display temperature to the nearest whole degree, such as 23°C. The displayed temperature changes by jumps of 1°C. As a result, temperatures between two whole degrees, such as 22.7°C, are not displayed.

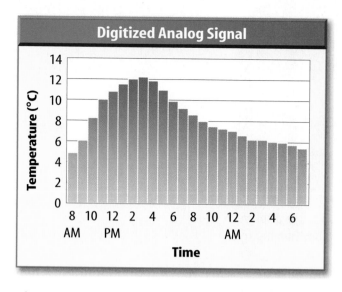

Making Digital Signals A smoothly varying analog signal can be converted to a digital signal. For example, suppose you wish to create a record of how the temperature outside changed over a day. One way to do this would be read an outdoor thermometer every hour and record the temperature and time. At the end of the day your temperature record would be a series of numbers. If you used these numbers to make a graph of the temperature record, it might look like the one shown in **Figure 3.** The temperature information shown by the graph changes in steps and is a digital signal.

Figure 3 A temperature record made by recording the temperature every hour changes in steps and is a digital signal.

Sampling an Analog Signal By recording the temperature every hour, you have sampled the smoothly varying outdoor temperature. When an analog signal is sampled, a value of the signal is read and recorded at some time interval, such as every hour or every second. An example is shown in **Figure 4.** As a result, a smoothly changing analog signal is converted to a series of numbers. This series of numbers is a digital signal.

The process of converting an analog signal to a digital signal is called digitization. The analog signal on a magnetic tape can be converted to a digital signal by sampling. In this way, a song can be represented by a series of numbers.

Figure 4 An analog signal can be converted to a digital signal. At a fixed time interval, the strength of the analog signal is measured and recorded. The resulting digital signal changes in steps.

Using Digital Signals It might seem that analog signals would be more useful than digital signals. After all, when an analog signal is converted to a digital signal, some information is lost. However, think about how analog and digital signals might be stored. Suppose a song that is stored as an analog signal on a small cassette tape were digitized and converted into a series of numbers. It might take millions of numbers to digitize a song, so how could these numbers be stored? As you will see later in this chapter, there is one electronic device that can store these numbers easily—a computer.

Once a digital signal is stored on a computer as a series of numbers, the computer can change these numbers using mathematical formulas. This process changes the signal and is called signal processing. For example, background noise can be removed from a digitized song using signal processing.

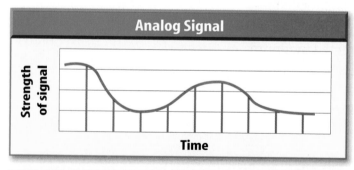

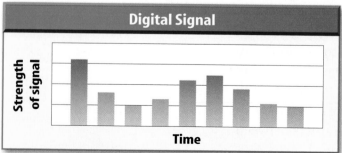

Early Television

Vacuum Tube

Modern Television

Electronic Devices

An electronic device, such as a calculator or a CD player, uses the information contained in electronic signals to do a job. For example, the job can be adding two numbers together or making sounds and images. The electronic signals are electric currents that flow through circuits in the electronic device. An electronic device, such as a calculator or a VCR, may contain hundreds or thousands of complex electric circuits.

Electronic Components The electric circuits in an electronic device usually contain electronic components. These electronic components are small devices that use the information in the electronic signals to control the flow of current in the circuits.

Early electronic devices, such as the early television shown in **Figure 5,** used electronic components called vacuum tubes, such as the one shown in the middle of **Figure 5,** to help create sounds and images. Vacuum tubes were bulky and generated a great deal of heat. As a result, early electronic devices used more electric power and were less dependable than those used today, such as the modern television shown in **Figure 5.** Today, televisions and radios no longer use vacuum tubes. Instead, they contain electronic components made from semiconductors.

Semiconductors

On the periodic table, the small number of elements found between the metals and nonmetals are called metalloids. Some metalloids, such as silicon and germanium, are semiconductors. A **semiconductor** is an element that is a poorer conductor of electricity than metals but a better conductor than nonmetals. However, semiconductors have a special property that ordinary conductors and insulators lack—their electrical conductivity can be controlled by adding impurities.

Figure 5 Because early televisions used vacuum tubes, they used more electrical power and were less reliable than their modern versions.

Science nline

Topic: Semiconductor Devices
Visit bookn.msscience.com for Web links to information about semiconductor devices.

Activity Choose one semiconductor device and write a paragraph explaining one way that it is used.

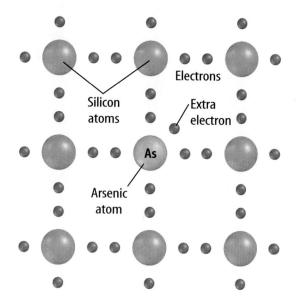

Figure 6 When arsenic atoms are added to a silicon crystal, they add extra electrons that are free to move about. This causes the electrical conductivity of the silicon crystal to increase.

Labels in figure: Silicon atoms, Electrons, Extra electron, As, Arsenic atom

INTEGRATE Chemistry

Adding Impurities Adding even a single atom of an element such as gallium or arsenic to a million silicon atoms significantly changes the conductivity. This process of adding impurities is called doping.

Doping can produce two different kinds of semiconductors. One type of semiconductor can be created by adding atoms like arsenic to a silicon crystal, as shown in **Figure 6.** Then the silicon crystal contains extra electrons. A semiconductor with extra electrons is an n-type semiconductor.

A p-type semiconductor is produced when atoms like gallium are added to a silicon crystal. Then the silicon crystal has fewer electrons than it had before. An n-type semiconductor can give, or donate, electrons and a p-type semiconductor can take, or accept, electrons.

✔ Reading Check *How are n-type and p-type semiconductors different?*

Solid-State Components

The two types of semiconductors can be put together to form electronic components that can control the flow of electric current in a circuit. Combinations of n-type and p-type semiconductors can form components that behave like switches that can be turned off and on. Other combinations can form components that can increase, or amplify, the change in an electric current or voltage. Electronic components that are made from combinations of semiconductors are called solid-state components. Diodes and transistors are examples of solid-state components that often are used in electric circuits.

Figure 7 Diodes like these allow current to flow in only one direction.

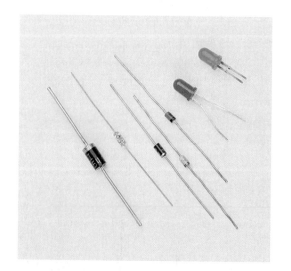

Diodes A **diode** is a solid-state component that, like a one-way street, allows current to flow only in one direction. In a diode, a p-type semiconductor is connected to an n-type semiconductor. Because an n-type semiconductor gives electrons and a p-type semiconductor accepts electrons, current can flow from the n-type to the p-type semiconductor, but not in the opposite direction. **Figure 7** shows common types of diodes. Diodes are useful for converting alternating current (AC) to direct current (DC). Recall that an alternating current constantly changes direction. When an alternating current reaches a diode, the diode allows the current to flow in only one direction. The result is direct current.

Transistors A **transistor** is a solid-state component that can be used to amplify signals in an electric circuit. A transistor also is used as an electronic switch. Electronic signals can cause a transistor to allow current to pass through it or to block the flow of current. **Figure 8** shows examples of transistors that are used in many electronic devices. Unlike a diode, a transistor is made from three layers of n-type and p-type semiconductor material sandwiched together.

Figure 8 Transistors such as these are used in electric circuits to amplify signals or to act as switches.

Integrated Circuits Personal computers usually contain millions of transistors, and would be many times larger if they used transistors the size of those shown in **Figure 8.** Instead, computers and other electronic devices use integrated circuits. An **integrated circuit** contains large numbers of interconnected solid-state components and is made from a single chip of semiconductor material such as silicon. An integrated circuit, like the one shown in **Figure 9,** may be smaller than 1 mm on each side and still can contain millions of transistors, diodes, and other components.

Figure 9 This tiny integrated circuit contains thousands of diodes and transistors.

section 1 review

Summary

Electronic Signals

- An electronic signal is a changing electric current that carries information.
- Analog electronic signals change continuously and digital electronic signals change in steps.
- An analog signal can be converted to a digital signal that is a series of numbers.

Solid-State Components

- Adding impurities to silicon can produce n-type semiconductors that donate electrons and p-type semiconductors that accept electrons.
- Solid-state components are electronic devices, such as diodes and transistors, made from n-type and p-type semiconductors.
- An integrated circuit contains a large number of solid-state components on a single semiconductor chip.

Self Check

1. **Explain** why the electric current that flows in a lamp is not an electronic signal.
2. **Describe** two advantages of using integrated circuits instead of vacuum tubes in electronic devices.
3. **Explain** why a digital signal can be stored on a computer.
4. **Compare and contrast** diodes and transistors.
5. **Think Critically** When an analog signal is sampled, what are the advantages and disadvantages of decreasing the time interval?

Applying Math

6. **Digital Signal** A song on a cassette tape is sampled and converted to a digital signal that is stored on a computer. The strength of the analog signal produced by the tape is sampled every 0.1 s. If the song is 3 min and 20 s long, how many numbers are in the digital signal stored on the computer?

Investigating Diodes

Diodes are found in most electronic devices. They are used to control the flow of electrons through a circuit. Electrons will flow through a diode in only one direction, from the n-type semiconductor to the p-type semiconductor. In this lab you will use a type of diode called an LED (light-emitting diode) to observe how a diode works.

▶ Real-World Question

How does electric current flow through a diode?

Goals
■ **Create** an electronic circuit.
■ **Observe** how an LED works.

Materials
light-emitting diode   D-cell battery and holder
lightbulb and holder   wire

Safety Precautions

▶ Procedure

1. Set up the circuit shown below. Record your observations. Then reverse the connections so each wire is connected to the other battery terminals. Record your observations.

2. Disconnect the wires from the lightbulb and attach one wire to each end of an LED. Observe whether the LED lights up when you connect the battery.

3. Reverse the connections on the LED so the current goes into the opposite end. Observe whether the LED lights up this time. Record your observation.

▶ Conclude and Apply

1. **Explain** why the bulb did or did not light up each time.

2. **Explain** why the LED did or did not light up each time.

3. **Describe** how the behavior of the lightbulb is different from that of the LED.

4. **Infer** which wire on the LED is connected to the n-type semiconductor and which is connected to the p-type semiconductor based on your observations.

𝒞ommunicating Your Data

Discuss your results with other students in your class. Did their LEDs behave in the same way? **For more help, refer to the Science Skill Handbook.**

1.5V Battery

Lightbulb

Computers

What are computers?

When was the last time you used a computer? Computers are found in libraries, grocery stores, banks, and gas stations. Computers seem to be everywhere. A computer is an electronic device that can carry out a set of instructions, or a program. By changing the program, the same computer can be made to do a different job.

Compared to today's desktop and laptop computers, the first electronic computers, like the one shown in **Figure 10,** were much bigger and slower. Several of the first electronic computers were built in the United States between 1946 and 1951. Solid-state components and the integrated circuit had not been developed yet. So these early computers contained thousands of vacuum tubes that used a great deal of electric power and produced large amounts of heat.

Computers became much smaller, faster, and more efficient after integrated circuits became available in the 1960s. Today, even a game system, like the one in **Figure 10,** can carry out many more operations each second than the early computers.

as you read

What You'll Learn
- **Describe** the different parts of a computer.
- **Compare** computer hardware with computer software.
- **Discuss** the different types of memory and storage in a computer.

Why It's Important
You can do more with computers if you understand how they work.

Review Vocabulary
laser: a device that produces a concentrated beam of light

New Vocabulary
- binary system
- random-access memory
- read-only memory
- computer software
- microprocessor

Figure 10 One of the first electronic computers was ENIAC, which was built in 1946 and weighed more than 30 tons. ENIAC could do 5,000 additions per second.

This handheld game system can do millions of operations per second.

Table 1 Combinations of Binary Digits	
Number of of Binary Digits	Possible Combinations
1	0 1
2	00 01 10 11
3	000 001 010 011 100 101 110 111

Computer Information

How does a computer display images, generate sounds, and manipulate numbers and words? Every piece of information that is stored in or used by a computer must be converted to a series of numbers. The words you write with a word processor, or the numbers in a spreadsheet are stored in the computer's memory as numbers. An image or a sound file also is stored as a series of numbers. Information stored in this way is sometimes called digital information.

Binary Numbers Imagine what it would be like if you had to communicate with just two words—on and off. Could you use these words to describe your favorite music or to read a book out loud? Communication with just two words seems impossible, but that's exactly what a computer does.

All the digital information in a computer is converted to a type of number that is expressed using only two digits—0 and 1. This type of number is called a binary (BI nuh ree) number. Each 0 or 1 is called a binary digit, or bit. Because this number system uses only two digits, it is called the **binary system,** or base-2 number system.

✔ **Reading Check** *Which digits are used in the binary system?*

Combining Binary Digits You might think that using only two digits would limit the amount of information you can represent. However, a small number of binary digits can be used to generate a large number of combinations, as shown in **Table 1.**

While one binary digit has only two possible combinations—0 or 1—there are four possible combinations for a group of two binary digits, as shown in **Table 1.** By using just one more binary digit the possible number of combinations is increased to eight. The number of combinations increases quickly as more binary digits are added to the group. For example, there are 65,536 combinations possible for a group of 16 binary digits.

Representing Information with Binary Digits

Combinations of binary digits can be used to represent information. For example, the English alphabet has 26 letters. Suppose each letter was represented by one combination of binary digits. To represent both lowercase and uppercase letters would require a total of 52 different combinations of binary digits. Would a group of five binary digits have enough possible combinations?

Representing Letters and Numbers A common system that is used by computers represents each letter, number, or other text character by eight binary digits, or one byte. There are 256 combinations possible for a group of eight binary digits. In this system, the letter "A" is represented by the byte 01000001, while the letter "a" is represented by the byte 01100001, and a question mark is represented by 00111111.

Computer Memory

Why are digital signals stored in a computer as binary numbers? A binary number is a series of bits that can have only one of two values—0 and 1. A switch, such as a light switch on a wall, can have two positions: on or off. A switch could be used to represent the two values of a bit. A switch in the "off" position could represent a 0, and a switch in the "on" position could represent a 1. **Table 2** shows how switches could be used to represent combinations of binary digits.

Table 2 Representing Binary Digits	
Binary Number	Switches
0000	
0001	
0010	
0011	
0100	
1010	

Applying Science

How much information can be stored?

Information can be stored in a computer's memory or in storage devices such as hard disks or CDs. The amount of information that can be stored is so large that special units, shown in the table on the right, are used. Desktop computers often have hard disks that can store many gigabytes of information. How much information can be stored in one gigabyte of storage?

Size of Information Storage Units	
Information Storage Unit	Number of Bytes
kilobyte	1,024
megabyte	1,048,576
gigabyte	1,073,741,824

Identifying the Problem

When words are stored on a computer, every letter, punctuation mark, and space between words is represented by one byte. A page of text, such as this page, might contain as many as 2,900 characters. So to store a page of text on a computer might require 2,900 bytes.

If you write a page of text using a word-processing program, more bytes might be needed to store the page. This is because when the page is stored, some word-processing programs include other information along with the text.

Solving the Problem

1. If it takes 2,900 bytes to store one page of text on a computer, how many pages can be stored in 1 gigabyte of storage?
2. Suppose a book contains 400 pages of text. How many books could be stored on a 1-gigabyte hard disk?
3. A CD can hold 650 megabytes of information. How many 400-page books could be stored on a CD?

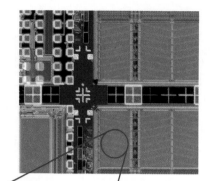

Figure 11 Computer memory is made of integrated circuits like this one. This integrated circuit can contain millions of microscopic circuits, shown here under high magnification.

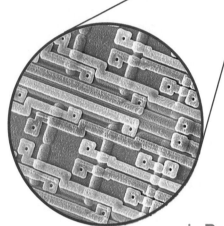

Storing Information The memory in a computer is an integrated circuit that contains millions of tiny electronic circuits, as shown in **Figure 11.** In the most commonly used type of computer memory, each circuit is able to store electric charge and can be either charged or uncharged. If the circuit is charged, it represents the bit 1 and if it is uncharged it represents the bit 0. Because computer memory contains millions of these circuits, it can store tremendous amounts of information using only the numbers 1 and 0.

What is your earliest memory? When you remember something from long ago, you use your long-term memory. On the other hand, when you work on a math problem, you may keep the numbers in your head long enough to find the answer. Like you, a computer has a long-term memory and a short-term memory that are used for different purposes.

Random-Access Memory A computer's **random-access memory,** or RAM, is short-term memory that stores documents, programs, and data while they are being used. Program instructions and data are temporarily stored in RAM while you are using a program or changing the data.

For example, a computer game is kept in RAM while you are playing it. If you are using a word-processing program to write a report, the report is temporarily held in RAM while you are working on it. Because information stored in RAM is lost when the computer is turned off, this type of memory cannot store anything that you want to use later.

The amount of RAM depends on the number of binary digits it can store. Recall that eight bits is called a byte. A megabyte is more than one million bytes. A computer that has 128 megabytes of memory can store more than 128 million bytes of information in its RAM, or nearly one billion bits.

Reading Check *What happens to information in RAM when the computer is turned off?*

Read-Only Memory Some information that is needed to enable the computer to operate is stored in its permanent memory. The computer can read this memory, but it cannot be changed. Memory that can't be changed and is permanently stored inside the computer is called **read-only memory,** or ROM. ROM is not lost when the computer is turned off.

Science Online

Topic: Computer Software
Visit bookn.msscience.com for Web links to information about types of computer software.

Activity Choose one type of software application and write a paragraph explaining why it is useful. Create a chart that summarizes what the software does.

Computer Programs

It's your mother's birthday and you decide to surprise her by baking a chocolate cake. You find a recipe for chocolate cake in a cookbook and follow the directions in the order the recipe tells you to. However, if the person who wrote the recipe left out any steps or put them in the wrong order, the cake probably will not turn out the way you expected. A computer program is like a recipe. A program is a series of instructions that tell the computer how to do a job. Unlike the recipe for a cake, some computer programs contain millions of instructions that tell the computer how to do many different jobs.

All the functions of a computer, such as displaying an image on the computer monitor or doing a math calculation, are controlled by programs. These instructions tell the computer how to add two numbers, how to display a word, or how to change an image on the monitor when you move a joystick. Many different programs can be stored in a computer's memory.

Computer Software When you type a report, play a video game, draw a picture, or look through an encyclopedia on a computer, you are using computer software. **Computer software** is any list of instructions for the computer. The instructions that are part of the software tell the computer what to display on the monitor. If you respond to what you see, for example by moving the mouse, the software instructions tell the computer how to respond to your action.

Computer Programming

The process of writing computer software is called computer programming. To write a computer program, you must decide what you want the computer to do, plan the best way to organize the instructions, write the instructions, and test the program to be sure it works. A person who writes computer programs is called a computer programmer. Computer programmers write software in computer languages such as Basic, C++, and Java.

Figure 12 shows part of a computer program. After the program is written, it is converted into binary digits to enable it to be stored in the computer's memory. Then the computer can carry out the program's instructions.

Mini LAB

Observing Memory

Procedure
1. Write a different five-digit number on six 3 × 5 cards.
2. Show a card to a partner for 3 s. Turn the card over and ask your partner to repeat the number. Repeat with two other cards.
3. Repeat this procedure with the last three cards, but wait 20 s before asking your partner to repeat each number.

Analysis
Is your partner's memory of the five-digit numbers more like computer RAM or ROM? Explain.

Try at Home

Figure 12 The text below is part of a computer program that directs the operation of a computer.

```
int request_dma(unsigned int dmanr, const char * device_id)
{
    if (dmanr > = MAX_DMA_CHANNELS)
        return -EINVAL;

    if (xchg(&dma_chan_busy[dmanr].lock, 1) != 0)
        return -EBUSY;

    dma_chan_busy[dmanr].device_id = device_id;

    /* old flag was 0, now contains 1 to indicate busy */
    return 0;
} /* request_dma */

void free_dma(unsigned int dmanr)
{
    if (dmanr > = MAX_DMA_CHANNELS) {
        printk("Trying to free DMA%d\n", dmanr);
        return;
    }
```

Computer Programmers
Computer software is written by computer programmers. It may take from a few hours to more than a year to write a program, and involve a single programmer or a team of programmers. A programmer usually must know several computer languages, such as COBOL, Java, and C++. Training in computer languages is required, and most jobs also require a college degree. Research to find the schools in your area that offer training as a computer programmer.

Computer Hardware

When you press a key on a computer's keyboard, a letter appears on the screen. This seems to occur all at once, but actually three steps are involved. In the first step, the computer receives information from an input device, such as a keyboard or mouse. For example, when you press a key on the keyboard, the computer receives and stores an electronic signal from the keyboard.

The next step is to process the input signal from the keyboard. This means to change the input signal into an electronic signal that can be understood by the computer monitor. The computer does this by following instructions contained in the programs stored in the computer's memory. The third step is to send the processed signal to the monitor.

All three steps can be carried out with a combination of hardware and software components. Computer hardware consists of input devices, output devices, storage devices, and integrated circuits for storing information. A keyboard and a mouse are examples of input devices, while a monitor, a printer, and loudspeakers are examples of output devices. Storage devices, such as floppy disks, hard disks, and CDs, are used to store information outside of the computer memory. A computer also contains a microprocessor that controls the computer hardware. Examples of computer hardware are shown in **Figure 13.**

Figure 13 Computer hardware includes input devices, output devices, and storage devices.

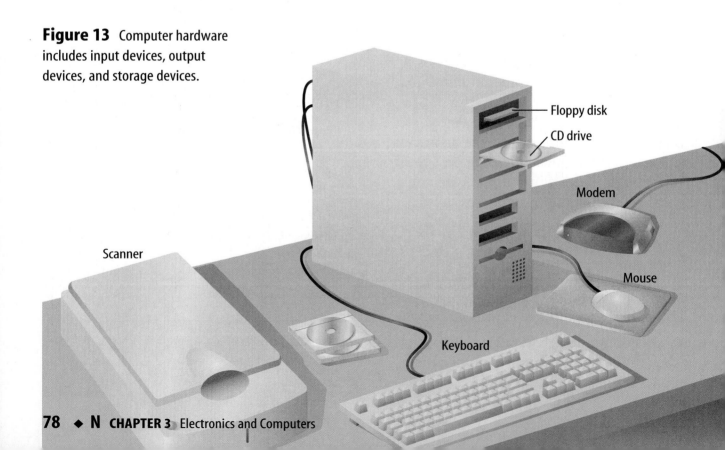

The Microprocessor Modern computers contain a microprocessor, like the one shown in **Figure 14,** that serves as the brain of the computer. A **microprocessor,** which is also called the central processing unit, or CPU, is an integrated circuit that controls the flow of information between different parts of the computer. A microprocessor can contain millions of interconnected transistors and other components. The microprocessor receives electronic signals from various parts of the computer, processes these signals, and sends electronic signals to other parts of the computer. For example, the microprocessor might tell the hard-disk drive to write data to the hard disk or the monitor to change the image on the screen. The microprocessor does this by carrying out instructions that are contained in computer programs stored in the computer's memory.

The microprocessor was developed in the late 1970s as the result of a process that made it possible to fit thousands of electronic components on a silicon chip. In the 1980s, the number of components on a silicon chip increased to hundreds of thousands. In the 1990s, microprocessors were developed that contained several million components on a single chip.

Figure 14 The pencil points to the microprocessor in the photo above. This microprocessor has dimensions of about one centimeter on a side, but contains millions of transistors and other solid-state components.

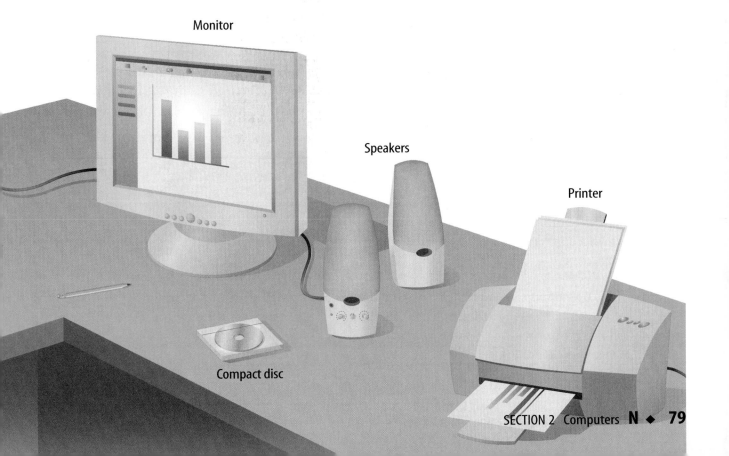

Monitor

Speakers

Printer

Compact disc

Storing Information

You have decided to type your homework assignment on a computer. The resulting paper is quite long and you make many changes to it each time you read it. How does the computer make it possible for you to store your information and make changes to it?

Both RAM and ROM are integrated circuits inside the computer. You might wonder, then, why other types of information storage are needed. Information stored in RAM is lost when the computer is turned off, and information stored in ROM can only be read—it can't be changed. If you want to store information that can be changed but isn't lost when the computer is off, you must store that information on a storage device, such as a disk. Several different types of disks are available.

Hard Disks A hard disk is a device that stores computer information magnetically. A hard disk is usually located inside a computer. **Figure 15** shows the inside of a hard disk, and **Figure 16** shows how a hard disk stores data. The hard disk contains one or more metal disks that have magnetic particles on one surface. When you save information on a hard disk, a device called a read/write head inside the disk drive changes the orientation of the magnetic particles on the disk's surface. Orientation in one direction represents 0 and orientation in the opposite direction represents 1. When a magnetized disk is read, the read/write head converts the digital information on the disk to pulses of electric current.

Information stored magnetically cannot be read by the computer as quickly as information stored on RAM and ROM. However, because the information on a hard disk is stored magnetically rather than with electronic switches like RAM, the information isn't lost when the computer is turned off.

Figure 15 A hard disk contains a disk or platter that is coated with magnetic particles. A read/write head moves over the surface of the disk.

Figure 16

Computers are useful because they can process large amounts of information quickly. Almost all desktop computers use a hard disk to store information. A hard disk is an electronic filing cabinet that can store enormous amounts of information and retrieve them quickly.

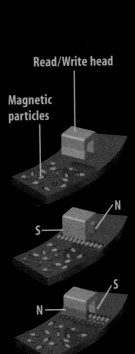

Hard disk unit

Read/Write head

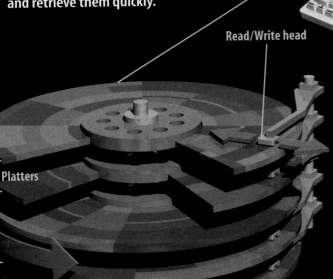

Platters

Read/Write head

Magnetic particles

A A hard-disk drive is made of a stack of aluminum disks, called platters, that are coated with a thin layer that contains magnetic particles. Like tiny compasses, these particles will line up along magnetic field lines. The hard disk also contains read/write heads that contain electromagnets. When the hard disk is turned on, the platters spin under the heads.

B To write information on the disk, a magnetic field is created around the head by an electric current. As the platter rotates past the head, this magnetic field causes the magnetic particles on the platter to line up in bands. One direction of the bands corresponds to the digital bit 0, the other to the digital bit 1.

C To read information on the disk, no current is sent to the heads. Instead, the magnetized bands create a changing current in the head as it passes over the platter. This current is the electronic signal that represents the needed information.

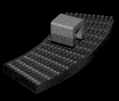

Floppy Disks Storing information on a hard disk is convenient, but sometimes you might want to store information that you can carry with you. The original storage device of this type was the floppy disk. A floppy disk is a thin, flexible, plastic disk. You might be confused by the term *floppy* if you have heard it used to describe disks that seem quite rigid. That is because you don't actually hold the floppy disk. Instead, you hold the harder plastic case in which the floppy disk is encased. Just as for a hard disk, the floppy disk is coated with a magnetic material that is magnetized and read by a read/write head. Floppy disks have lower storage capacity than hard disks. Also, compared to hard disks, information is read from and written to floppy disks much more slowly.

Optical Disks An optical storage disk, such as a CD, is a thin, plastic disk that has information digitally stored on it. The disk contains a series of microscopic pits and flat spots as shown on the left in **Figure 17.** A tiny laser beam shines on the surface of the disk. The information on the disk is read by measuring the intensity of the laser light reflected from the surface of the disk. This intensity will depend on whether the laser beam strikes a pit or a flat spot. The original optical storage disks, laser discs, CD-ROMs, and DVD-ROMs, were read-only. Several of these are shown on the right in **Figure 17.** However, CD-RW disks can be erased and rewritten many times. Information is written by a CD burner that causes a metal alloy in the disk to change form when heated by a laser. When the disk is read, the intensity of reflected laser light depends on which form of the alloy the beam strikes.

Science nline

Topic: Optical Disks

Visit bookn.msscience.com for Web links to information about storing data on optical disks.

Activity Make a table that shows the similarities and differences between CDs and DVDs.

Figure 17 An optical storage disk stores information that is read by a laser.

Explain *the difference between a read-only disk and a reusable disk.*

Information is stored on an optical disk by a series of pits and flat spots, representing a binary 1 or 0.

CDs, laser disks, and DVDs are all examples of optical storage disks.

Computer Networks

People can communicate using a computer if it is part of a computer network. A computer network is two or more computers that are connected to share files or other information. The computers might be linked by cables, telephone lines, or radio signals.

The Internet is a collection of computer networks from all over the world. The Internet is linked together by cable or satellite. The Internet itself has no information. No documents or files exist on the Internet, but you can use the Internet to access a tremendous amount of information by linking to other computers.

The World Wide Web is part of the Internet. The World Wide Web is the ever-changing collection of information (text, graphics, audio, and video) on computers all over the world. The computers that store these documents are called servers. When you connect with a server through the Internet, you can view any of the Web documents that are stored there, like the Web page shown in **Figure 18.** A particular collection of information that is stored in one place is known as a Web site.

Figure 18 When you connect to the Internet, you can be linked with other computers that are part of the World Wide Web. Then you can have access to the information stored at millions of Web sites.

section 2 review

Summary

Computer Information

- A binary digit can be a 0 or a 1.
- Computers store information as groups of binary digits.
- Computers use tiny electronic circuits to represent binary digits and store information.

Computer Software and Hardware

- Computer software and computer programs are lists of instructions for a computer.
- Computer programs are written in special computer languages.
- Computer hardware, such as keyboards and hard disks, is controlled by a microprocessor.

Storing Information

- Hard disks and floppy disks store information on disks coated with magnetic particles.
- Optical disks store information as a series of pits and flat spots that is read by a laser.

Self Check

1. **Explain** why the binary number system is used for storing information in computers.
2. **Compare and contrast** the Internet and the World Wide Web.
3. **Describe** what a microprocessor does with the signals it receives from various parts of a computer.
4. **Compare and contrast** three different computer information storage devices.
5. **Think Critically** Why can't computer information be stored only in RAM and ROM, making storage devices such as hard disks and optical disks unnecessary?

Applying Skills

6. **Make a Concept Map** Develop a spider map about computers. Include the following terms in your spider map: *keyboard, monitor, microprocessor, software, printer, RAM, ROM, floppy disk, hard disk, CD, Internet,* and *World Wide Web.*

Use the Internet

Does your computer have a virus?

Goals

- **Understand** what a computer virus is.
- **Identify** different types of computer viruses.
- **Describe** how a computer virus is spread.
- **Create** a plan for protecting electronic files and computers from computer viruses.

Data Source

ScienceOnline

Visit **bookn.msscience.com/ internet_lab** to get more information about computer viruses and for data collected by other students.

▶ Real-World Question

The Internet has provided many ways to share information and become connected with people near and far. People can communicate ideas and information quickly and easily. Unfortunately, some people use computers and the Internet as an opportunity to create and spread computer viruses. Many new viruses are created each year that can damage information and programs on a computer. Viruses create problems for computers in homes and schools. Computer problems caused by viruses can be costly for business and government computers, as well. How can acquiring and transmitting computer viruses be prevented?

▶ Make a Plan

People share information and ideas by exchanging electronic files with one another. Perhaps you send email to your friends and family. Many people send word processing or spreadsheet files to friends and associates. What happens if a computer file is infected with a virus? How is that virus spread among different users? How can you protect your computer and your information from being attacked by a virus?

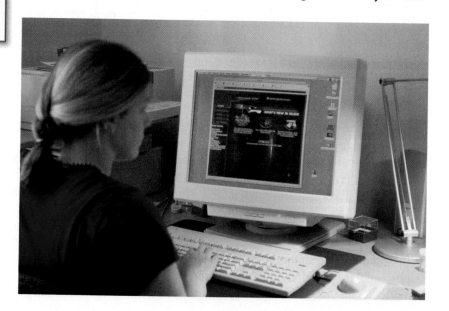

▶ Follow Your Plan

1. Do research to find out what a computer virus is and the difference between various types of viruses. Also research the ways that a computer virus can damage computer files and programs.

2. After you know what a computer virus is, make a list of different types of viruses and how they are passed from computer to computer. For example, some viruses can be passed through attachments to email. Others can be passed by sharing spreadsheet files. Be specific about how a virus is passed.

3. Discover how you can protect yourself from viruses that attack your computer. Make a list of steps to follow to avoid infection.

4. Make sure your teacher approves your plan before you start.

5. Visit the link below to post your data.

▶ Analyze Your Data

1. **Explain** how computer viruses are transferred from one computer to another.

2. **Explain** how you can prevent your computer from becoming infected by a virus.

3. **Explain** how you can prevent other people from getting computer viruses.

4. **Describe** the different ways computer viruses can damage computer files and programs.

▶ Conclude and Apply

1. **List** five to eight steps a computer user should follow to prevent getting a computer virus or passing a computer virus to another computer.

2. **Discuss** how antivirus software can keep viruses from spreading. Could antivirus software always prevent you from getting a computer virus? Why or why not?

Communicating Your Data

Find this lab using the link below. Post your data in the table that is provided. **Compare** your data on types of viruses and how they infect computers with that of other students.

Science Online

bookn.msscience.com/internet_lab

E-Lectrifying *E*-Books

Here's a look at how computers and the Internet are changing what—and how—you read

In recent years, people have been using their computers to order books from online bookstores. That's no big deal. What might become a big deal is the ordering of electronic books—books that you download to your own computer and read on the screen or print out to read later. Some famous authors are writing books just for that purpose. Some of the books are published only online—you can't find them anywhere else.

Many other Web sites, however, are selling any book anybody wishes to write—including students like you. In fact, you could start your own online bookstore with your own stories and reports. It will be up to readers to pick and choose what's good from the huge number of e-books that will be on the Web.

Curling Up with a Good Disk

Downloading books to your home computer is just one way to get an e-book. You can also buy versions of books to read on hand-held devices that are about the size of a paperback book. With one device, the books come on CD-ROM disks. With another, the books download to the device over a modem.

Current e-book devices are expensive, heavy, and awkward, and the number of books you can get for them is small. But if improvements

All of these stacked books can fit into one e-book.

come quickly, it might not be long before you check out of the library with a pocketful of disks instead of a heavy armload of books!

Will Traditional Books Disappear?

Most people think that the traditional printed book will never disappear. Publishers will still be printing books on paper with soft and hard covers. But publishers also predict there will be more and more kinds of formats for books. E-books, for example, might be best for interactive works that blend video, sound, and words the way many Web sites already do. For example, an e-book biography might allow the reader to click on photos and videos of the subject, and even provide links to other sources of information.

Interview Talk to a bookstore employee to find out how book publishing and selling has changed in the last five years. Can he or she predict how people will read books in the future? Report to the class.

Sciencenline

For more information, visit bookn.msscience.com/time

Reviewing Main Ideas

Section 1 Electronics

1. A changing electric current used to carry information is an electronic signal. Electronic signals can be either analog or digital.

2. Semiconductor elements, such as silicon and germanium, conduct electricity better than nonmetals but not as well as metals. If a small amount of some impurities is added to a semiconductor, its conductivity can be controlled.

3. Diodes and transistors are solid-state components. Diodes allow current to flow in one direction only. Transistors are used as switches or amplifiers.

Section 2 Computers

1. The binary system consists of two digits, 0 and 1. Switches within electronic devices such as computers can store information by turning on (1) and off (0).

2. Electronic memory within a computer can be random-access (RAM) or read-only (ROM).

3. Computer hardware consists of the physical parts of a computer. Computer software is a list of instructions for a computer.

4. A microprocessor is a complex integrated circuit that receives signals from various parts of the computer, processes these signals, and then sends instructions to various parts of the computer.

5. Floppy disks, hard disks, and optical disks are types of computer information storage devices.

6. The Internet is a collection of linked computer networks from all over the world. The World Wide Web is part of the Internet.

Visualizing Main Ideas

Copy and complete the following concept map on computers.

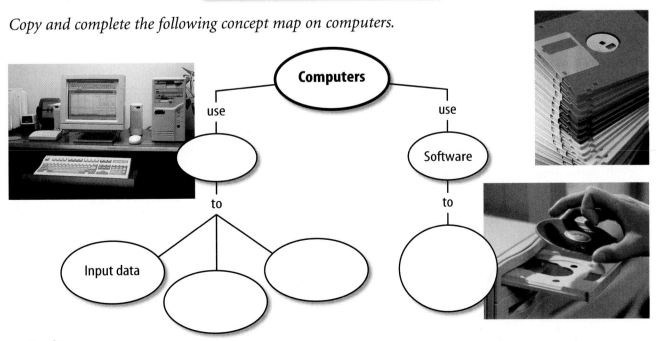

Using Vocabulary

analog signal p.66
binary system p.74
computer software p.77
digital signal p.67
diode p.70
electronic signal p.66
integrated circuit p.71

microprocessor p.79
random access
memory p.76
read-only memory p.76
semiconductor p.69
transistor p.71

Fill in the blanks with the correct vocabulary word or words.

1. _____ is a base-2 number system.

2. A(n) _____ can change AC current to DC current.

3. A(n) _____ is made from a single piece of semiconductor material and can contain thousands of solid-state components.

4. The information in a computer's _____ changes each time the computer is used.

5. An electronic device that can be used as a switch or to amplify electronic signals is a(n) _____.

6. A(n) _____ is also called a CPU.

7. An electronic signal that varies smoothly with time is a(n) _____.

Checking Concepts

Choose the word or phrase that best answers the question.

8. Which of the following best describes integrated circuits?
   A) They can be read with a laser.
   B) They use vacuum tubes as transistors and diodes.
   C) They contain pits and flat areas.
   D) They can be small and contain a large number of solid-state components.

9. Which type of elements are semiconductors?
   A) metals   C) metalloids
   B) nonmetals   D) gases

10. How is a digital signal different from an analog signal?
    A) It uses electric current.
    B) It varies continuously.
    C) It changes in steps.
    D) It is used as a switch.

11. Which of the following uses magnetic materials to store digital information.
    A) DVD   C) RAM
    B) hard disk   D) compact disk

12. Which part of a computer carries out the instructions contained in computer programs and software?
    A) RAM   C) hard disk
    B) ROM   D) microprocessor

13. Which type of computer memory is used when a computer is first turned on?
    A) ROM   C) DVD
    B) RAM   D) floppy disk

14. The instructions contained in a computer program are stored in which type of computer memory while the program is being used?
    A) ROM   C) CD
    B) RAM   D) floppy disk

Use the figure below to answer question 15.

15. What binary number is represented by the positions of the switches?
    A) 1110   C) 0101
    B) 0010   D) 0001

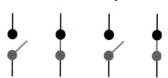

Thinking Critically

16. **Compare and contrast** an analog device and a digital device.

17. **Make and Use Tables** Copy and complete the following table that describes solid-state components.

Solid-State Components		
Component	Description	Use
Diode		
Transistor		
Integrated circuit		

18. **Explain** why the binary number system is used to store digital information in computers, instead of the decimal number system you use every day.

19. **Concept Map** Copy and complete the following events-chain map showing the sequence of events that occurs when a computer mouse is moved.

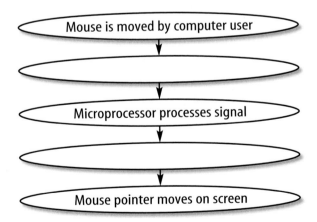

Mouse is moved by computer user

↓

↓

Microprocessor processes signal

↓

↓

Mouse pointer moves on screen

20. **Discuss** how the development of solid-state components and integrated circuits affected devices such as TVs and computers.

21. **Make a table** to classify the different types of internal and external computer memory and storage.

Performance Activities

22. **Make a Poster** Microprocessors continue to be developed that are more complex and contain an ever-increasing number of solid state components. Visit **msscience.com** for links to information about different microprocessors and how they have changed. Make a poster that summarizes what you have learned.

Applying Math

Students at a middle school researched the storage capacity of different computer storage devices. The information is summarized in the table below. The storage capacity is listed in units of gigabytes. A gigabyte is 1,074,000,000 bytes.

Use the table below to answer questions 23–25.

Computer Storage Devices	
Device	Capacity (Gb)
Floppy disk	0.00144
Compact disc	0.650
DVD	4.7
Hard Disk A	8.60
Hard Disk B	120.2

23. **Music Files Storage** In a certain format, to store 1 min of music as a digital signal requires 10,584,000 bytes. How many minutes of music in this format can be stored on the compact disc?

24. **Digital Pictures Storage** A certain digital camera produces digital images that require 921,600 bytes to store. How many of these images could be stored on hard disk A?

25. **Documents Storage** Seven documents produced by word processing software are stored on a floppy disk. If there are 40,000 bytes of storage still available on the disk, what is the average amount of storage used by each of the documents?

Part 1 | Multiple Choice

Record your answers on the answer sheet provided by your teacher or on a sheet of paper.

1. What kinds of materials are used to make solid-state components?
   A. semiconductors
   B. superconductors
   C. conductors
   D. insulators

2. Which of the following are not contained in integrated circuits?
   A. semiconductors   C. diodes
   B. vacuum tubes   D. transistors

Use the table below to answer questions 3 and 4.

Number of Binary Digit Combinations	
Number of Binary Digits	**Total Number of Combinations**
1	2
2	4
3	8
4	?
5	32

3. Which of the following is the total number of combinations of four binary digits?
   A. 64   C. 32
   B. 16   D. 8

4. Based on the data table, which of the following is the total number of combinations of six binary digits?
   A. 64   C. 32
   B. 16   D. 8

5. Which of the following best describes computer software?
   A. It is a type of temporary storage.
   B. It is a list of instructions.
   C. It contains analog information.
   D. It cannot be stored magnetically.

6. Which of the following is a computer input device?
   A. printer   C. monitor
   B. loudspeakers   D. keyboard

7. Which of the following is an optical storage device?
   A. hard disk   C. floppy disk
   B. RAM   D. CD

8. Where are elements that are semiconductors located on the periodic table?
   A. between metals and nonmetals
   B. on the right column
   C. on the left column
   D. at the bottom

9. Which of the following is not an electronic device?
   A. calculator   C. CD player
   B. television   D. light bulb

Use the figure below to answer question 10.

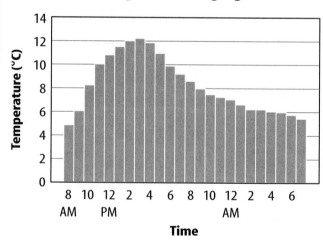

Digitized Analog Signal

10. Which of the following is a process that produces this digital signal from an analog signal?
    A. doping   C. switching
    B. programming   D. sampling

Part 2 | Short Response/Grid In

Record your answers on the answer sheet provided by your teacher or on a sheet of paper.

11. What is the difference between a binary 1 or 0 that is stored on a platter of a hard disk?

Use the figure below to answer questions 12 and 13.

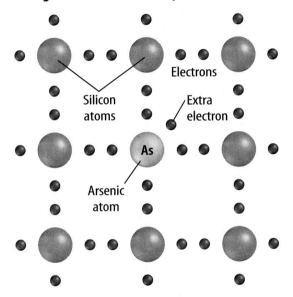

12. Describe the type of semiconductor material that is shown in the figure above.

13. Explain why atoms of other elements are added to a semiconductor material.

14. In a diode, which way does current flow between the n-type semiconductor and the p-type semiconductor?

15. Identify the process that enables background noise to be removed from a digitized music file that is stored on a hard disk.

Test-Taking Tip

Some Questions Have Qualifiers Look for qualifiers in a question. Such questions are not looking for absolute answers. Qualifiers could be words such as *most likely, most common,* or *least common.*

Part 3 | Open Ended

Record your answers on a sheet of paper.

16. A barograph is a device that measures and records air pressure. A barograph contains a pen that moves up and down as the pressure changes. The pen continuously draws a line on paper attached to a drum that slowly rotates. Infer whether the barograph is an analog or digital device. Explain.

17. Classify each of the following as an input, output, or storages device: monitor, keyboard, printer, hard disk. Explain the function of each device if you are using a word processing program to write a report.

18. Compare and contrast a floppy disk with an optical disk, such as a CD.

19. Explain why electric circuits that can be charged or uncharged are used as the components of computer memory.

Use the figure below to answer questions 20 and 21.

20. Describe how the read/write heads write information on a platter.

21. Describe how the read/write heads read information that is stored on the platters.

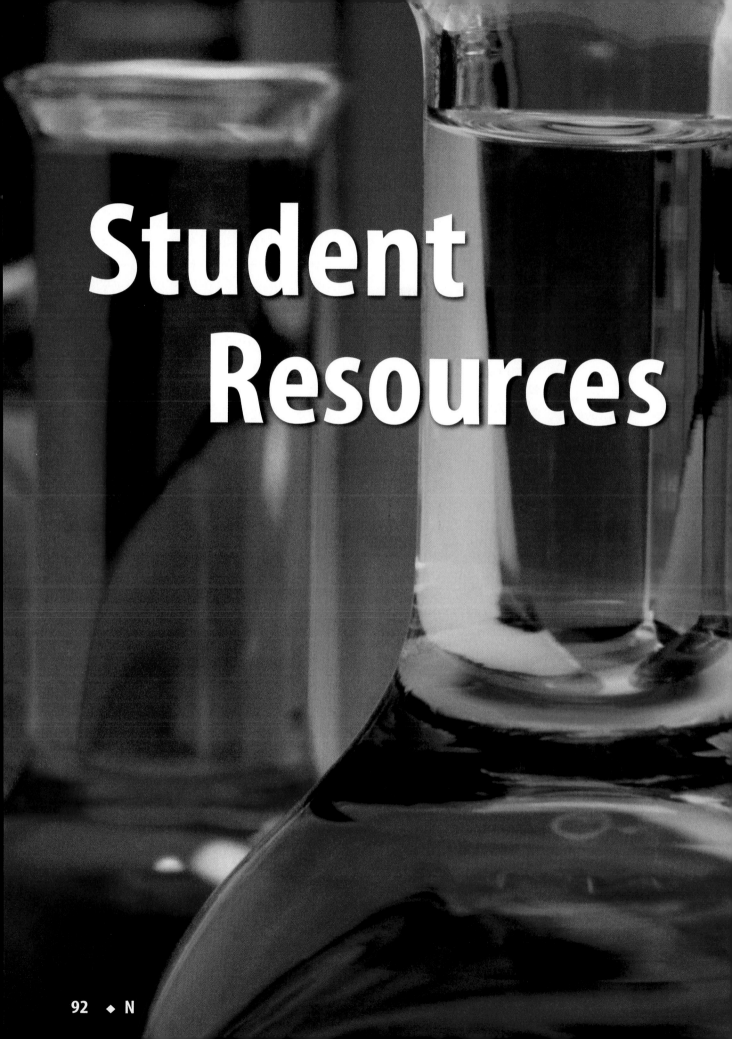

Student Resources

CONTENTS

Scientific Methods

Scientists use an orderly approach called the scientific method to solve problems. This includes organizing and recording data so others can understand them. Scientists use many variations in this method when they solve problems.

Identify a Question

The first step in a scientific investigation or experiment is to identify a question to be answered or a problem to be solved. For example, you might ask which gasoline is the most efficient.

Gather and Organize Information

After you have identified your question, begin gathering and organizing information. There are many ways to gather information, such as researching in a library, interviewing those knowledgeable about the subject, testing and working in the laboratory and field. Fieldwork is investigations and observations done outside of a laboratory.

Researching Information Before moving in a new direction, it is important to gather the information that already is known about the subject. Start by asking yourself questions to determine exactly what you need to know. Then you will look for the information in various reference sources, like the student is doing in **Figure 1.** Some sources may include textbooks, encyclopedias, government documents, professional journals, science magazines, and the Internet. Always list the sources of your information.

Figure 1 The Internet can be a valuable research tool.

Evaluate Sources of Information Not all sources of information are reliable. You should evaluate all of your sources of information, and use only those you know to be dependable. For example, if you are researching ways to make homes more energy efficient, a site written by the U.S. Department of Energy would be more reliable than a site written by a company that is trying to sell a new type of weatherproofing material. Also, remember that research always is changing. Consult the most current resources available to you. For example, a 1985 resource about saving energy would not reflect the most recent findings.

Sometimes scientists use data that they did not collect themselves, or conclusions drawn by other researchers. This data must be evaluated carefully. Ask questions about how the data were obtained, if the investigation was carried out properly, and if it has been duplicated exactly with the same results. Would you reach the same conclusion from the data? Only when you have confidence in the data can you believe it is true and feel comfortable using it.

Interpret Scientific Illustrations As you research a topic in science, you will see drawings, diagrams, and photographs to help you understand what you read. Some illustrations are included to help you understand an idea that you can't see easily by yourself, like the tiny particles in an atom in **Figure 2.** A drawing helps many people to remember details more easily and provides examples that clarify difficult concepts or give additional information about the topic you are studying. Most illustrations have labels or a caption to identify or to provide more information.

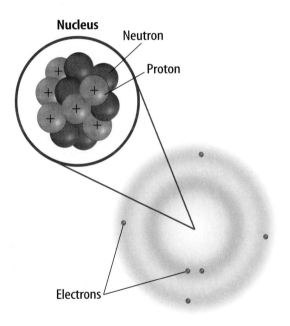

Figure 2 This drawing shows an atom of carbon with its six protons, six neutrons, and six electrons.

Concept Maps One way to organize data is to draw a diagram that shows relationships among ideas (or concepts). A concept map can help make the meanings of ideas and terms more clear, and help you understand and remember what you are studying. Concept maps are useful for breaking large concepts down into smaller parts, making learning easier.

Network Tree A type of concept map that not only shows a relationship, but how the concepts are related is a network tree, shown in **Figure 3.** In a network tree, the words are written in the ovals, while the description of the type of relationship is written across the connecting lines.

When constructing a network tree, write down the topic and all major topics on separate pieces of paper or notecards. Then arrange them in order from general to specific. Branch the related concepts from the major concept and describe the relationship on the connecting line. Continue to more specific concepts until finished.

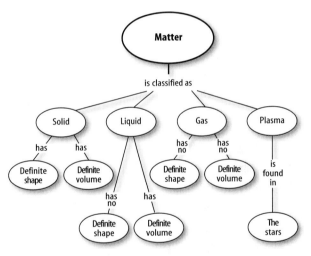

Figure 3 A network tree shows how concepts or objects are related.

Events Chain Another type of concept map is an events chain. Sometimes called a flow chart, it models the order or sequence of items. An events chain can be used to describe a sequence of events, the steps in a procedure, or the stages of a process.

When making an events chain, first find the one event that starts the chain. This event is called the initiating event. Then, find the next event and continue until the outcome is reached, as shown in **Figure 4.**

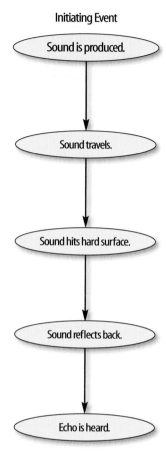

Initiating Event

Sound is produced.

Sound travels.

Sound hits hard surface.

Sound reflects back.

Echo is heard.

Figure 4 Events-chain concept maps show the order of steps in a process or event. This concept map shows how a sound makes an echo.

Cycle Map A specific type of events chain is a cycle map. It is used when the series of events do not produce a final outcome, but instead relate back to the beginning event, such as in **Figure 5.** Therefore, the cycle repeats itself.

To make a cycle map, first decide what event is the beginning event. This is also called the initiating event. Then list the next events in the order that they occur, with the last event relating back to the initiating event. Words can be written between the events that describe what happens from one event to the next. The number of events in a cycle map can vary, but usually contain three or more events.

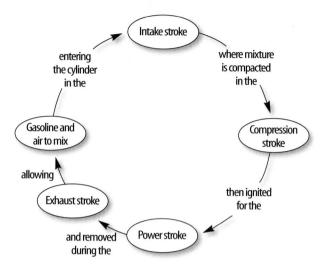

Figure 5 A cycle map shows events that occur in a cycle.

Spider Map A type of concept map that you can use for brainstorming is the spider map. When you have a central idea, you might find that you have a jumble of ideas that relate to it but are not necessarily clearly related to each other. The spider map on sound in **Figure 6** shows that if you write these ideas outside the main concept, then you can begin to separate and group unrelated terms so they become more useful.

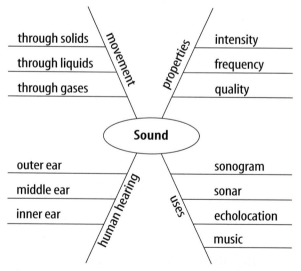

Figure 6 A spider map allows you to list ideas that relate to a central topic but not necessarily to one another.

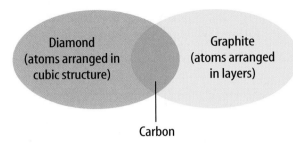

Figure 7 This Venn diagram compares and contrasts two substances made from carbon.

Venn Diagram To illustrate how two subjects compare and contrast you can use a Venn diagram. You can see the characteristics that the subjects have in common and those that they do not, shown in **Figure 7.**

To create a Venn diagram, draw two overlapping ovals that that are big enough to write in. List the characteristics unique to one subject in one oval, and the characteristics of the other subject in the other oval. The characteristics in common are listed in the overlapping section.

Make and Use Tables One way to organize information so it is easier to understand is to use a table. Tables can contain numbers, words, or both.

To make a table, list the items to be compared in the first column and the characteristics to be compared in the first row. The title should clearly indicate the content of the table, and the column or row heads should be clear. Notice that in **Table 1** the units are included.

Table 1 Recyclables Collected During Week			
Day of Week	**Paper (kg)**	**Aluminum (kg)**	**Glass (kg)**
Monday	5.0	4.0	12.0
Wednesday	4.0	1.0	10.0
Friday	2.5	2.0	10.0

Make a Model One way to help you better understand the parts of a structure, the way a process works, or to show things too large or small for viewing is to make a model. For example, an atomic model made of a plastic-ball nucleus and pipe-cleaner electron shells can help you visualize how the parts of an atom relate to each other. Other types of models can by devised on a computer or represented by equations.

Form a Hypothesis

A possible explanation based on previous knowledge and observations is called a hypothesis. After researching gasoline types and recalling previous experiences in your family's car you form a hypothesis—our car runs more efficiently because we use premium gasoline. To be valid, a hypothesis has to be something you can test by using an investigation.

Predict When you apply a hypothesis to a specific situation, you predict something about that situation. A prediction makes a statement in advance, based on prior observation, experience, or scientific reasoning. People use predictions to make everyday decisions. Scientists test predictions by performing investigations. Based on previous observations and experiences, you might form a prediction that cars are more efficient with premium gasoline. The prediction can be tested in an investigation.

Design an Experiment A scientist needs to make many decisions before beginning an investigation. Some of these include: how to carry out the investigation, what steps to follow, how to record the data, and how the investigation will answer the question. It also is important to address any safety concerns.

Test the Hypothesis

Now that you have formed your hypothesis, you need to test it. Using an investigation, you will make observations and collect data, or information. This data might either support or not support your hypothesis. Scientists collect and organize data as numbers and descriptions.

Follow a Procedure In order to know what materials to use, as well as how and in what order to use them, you must follow a procedure. **Figure 8** shows a procedure you might follow to test your hypothesis.

> **Procedure**
> 1. Use regular gasoline for two weeks.
> 2. Record the number of kilometers between fill-ups and the amount of gasoline used.
> 3. Switch to premium gasoline for two weeks.
> 4. Record the number of kilometers between fill-ups and the amount of gasoline used.

Figure 8 A procedure tells you what to do step by step.

Identify and Manipulate Variables and Controls In any experiment, it is important to keep everything the same except for the item you are testing. The one factor you change is called the independent variable. The change that results is the dependent variable. Make sure you have only one independent variable, to assure yourself of the cause of the changes you observe in the dependent variable. For example, in your gasoline experiment the type of fuel is the independent variable. The dependent variable is the efficiency.

Many experiments also have a control—an individual instance or experimental subject for which the independent variable is not changed. You can then compare the test results to the control results. To design a control you can have two cars of the same type. The control car uses regular gasoline for four weeks. After you are done with the test, you can compare the experimental results to the control results.

Collect Data

Whether you are carrying out an investigation or a short observational experiment, you will collect data, as shown in **Figure 9.** Scientists collect data as numbers and descriptions and organize it in specific ways.

Observe Scientists observe items and events, then record what they see. When they use only words to describe an observation, it is called qualitative data. Scientists' observations also can describe how much there is of something. These observations use numbers, as well as words, in the description and are called quantitative data. For example, if a sample of the element gold is described as being "shiny and very dense" the data are qualitative. Quantitative data on this sample of gold might include "a mass of 30 g and a density of 19.3 g/cm^3."

Figure 9 Collecting data is one way to gather information directly.

Figure 10 Record data neatly and clearly so it is easy to understand.

When you make observations you should examine the entire object or situation first, and then look carefully for details. It is important to record observations accurately and completely. Always record your notes immediately as you make them, so you do not miss details or make a mistake when recording results from memory. Never put unidentified observations on scraps of paper. Instead they should be recorded in a notebook, like the one in **Figure 10.** Write your data neatly so you can easily read it later. At each point in the experiment, record your observations and label them. That way, you will not have to determine what the figures mean when you look at your notes later. Set up any tables that you will need to use ahead of time, so you can record any observations right away. Remember to avoid bias when collecting data by not including personal thoughts when you record observations. Record only what you observe.

Estimate Scientific work also involves estimating. To estimate is to make a judgment about the size or the number of something without measuring or counting. This is important when the number or size of an object or population is too large or too difficult to accurately count or measure.

Sample Scientists may use a sample or a portion of the total number as a type of estimation. To sample is to take a small, representative portion of the objects or organisms of a population for research. By making careful observations or manipulating variables within that portion of the group, information is discovered and conclusions are drawn that might apply to the whole population. A poorly chosen sample can be unrepresentative of the whole. If you were trying to determine the rainfall in an area, it would not be best to take a rainfall sample from under a tree.

Measure You use measurements everyday. Scientists also take measurements when collecting data. When taking measurements, it is important to know how to use measuring tools properly. Accuracy also is important.

Length To measure length, the distance between two points, scientists use meters. Smaller measurements might be measured in centimeters or millimeters.

Length is measured using a metric ruler or meter stick. When using a metric ruler, line up the 0-cm mark with the end of the object being measured and read the number of the unit where the object ends. Look at the metric ruler shown in **Figure 11.** The centimeter lines are the long, numbered lines, and the shorter lines are millimeter lines. In this instance, the length would be 4.50 cm.

Figure 11 This metric ruler has centimeter and millimeter divisions.

Mass The SI unit for mass is the kilogram (kg). Scientists can measure mass using units formed by adding metric prefixes to the unit gram (g), such as milligram (mg). To measure mass, you might use a triple-beam balance similar to the one shown in **Figure 12.** The balance has a pan on one side and a set of beams on the other side. Each beam has a rider that slides on the beam.

When using a triple-beam balance, place an object on the pan. Slide the largest rider along its beam until the pointer drops below zero. Then move it back one notch. Repeat the process for each rider proceeding from the larger to smaller until the pointer swings an equal distance above and below the zero point. Sum the masses on each beam to find the mass of the object. Move all riders back to zero when finished.

Instead of putting materials directly on the balance, scientists often take a tare of a container. A tare is the mass of a container into which objects or substances are placed for measuring their masses. To mass objects or substances, find the mass of a clean container. Remove the container from the pan, and place the object or substances in the container. Find the mass of the container with the materials in it. Subtract the mass of the empty container from the mass of the filled container to find the mass of the materials you are using.

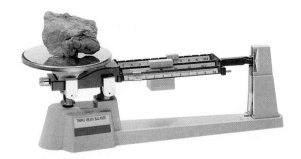

Figure 12 A triple-beam balance is used to determine the mass of an object.

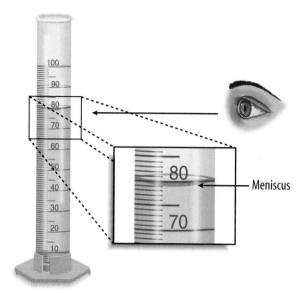

Figure 13 Graduated cylinders measure liquid volume.

Liquid Volume To measure liquids, the unit used is the liter. When a smaller unit is needed, scientists might use a milliliter. Because a milliliter takes up the volume of a cube measuring 1 cm on each side it also can be called a cubic centimeter (cm^3 = cm × cm × cm).

You can use beakers and graduated cylinders to measure liquid volume. A graduated cylinder, shown in **Figure 13,** is marked from bottom to top in milliliters. In lab, you might use a 10-mL graduated cylinder or a 100-mL graduated cylinder. When measuring liquids, notice that the liquid has a curved surface. Look at the surface at eye level, and measure the bottom of the curve. This is called the meniscus. The graduated cylinder in **Figure 13** contains 79.0 mL, or 79.0 cm^3, of a liquid.

Temperature Scientists often measure temperature using the Celsius scale. Pure water has a freezing point of 0°C and boiling point of 100°C. The unit of measurement is degrees Celsius. Two other scales often used are the Fahrenheit and Kelvin scales.

Analyze the Data

To determine the meaning of your observations and investigation results, you will need to look for patterns in the data. Then you must think critically to determine what the data mean. Scientists use several approaches when they analyze the data they have collected and recorded. Each approach is useful for identifying specific patterns.

Interpret Data The word *interpret* means "to explain the meaning of something." When analyzing data from an experiment, try to find out what the data show. Identify the control group and the test group to see whether or not changes in the independent variable have had an effect. Look for differences in the dependent variable between the control and test groups.

Classify Sorting objects or events into groups based on common features is called classifying. When classifying, first observe the objects or events to be classified. Then select one feature that is shared by some members in the group, but not by all. Place those members that share that feature in a subgroup. You can classify members into smaller and smaller subgroups based on characteristics. Remember that when you classify, you are grouping objects or events for a purpose. Keep your purpose in mind as you select the features to form groups and subgroups.

Compare and Contrast Observations can be analyzed by noting the similarities and differences between two more objects or events that you observe. When you look at objects or events to see how they are similar, you are comparing them. Contrasting is looking for differences in objects or events.

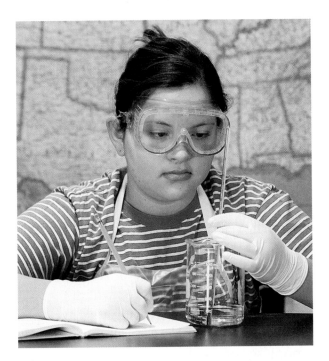

Figure 14 A thermometer measures the temperature of an object.

Scientists use a thermometer to measure temperature. Most thermometers in a laboratory are glass tubes with a bulb at the bottom end containing a liquid such as colored alcohol. The liquid rises or falls with a change in temperature. To read a glass thermometer like the thermometer in **Figure 14,** rotate it slowly until a red line appears. Read the temperature where the red line ends.

Form Operational Definitions An operational definition defines an object by how it functions, works, or behaves. For example, when you are playing hide and seek and a tree is home base, you have created an operational definition for a tree.

Objects can have more than one operational definition. For example, a ruler can be defined as a tool that measures the length of an object (how it is used). It can also be a tool with a series of marks used as a standard when measuring (how it works).

Recognize Cause and Effect A cause is a reason for an action or condition. The effect is that action or condition. When two events happen together, it is not necessarily true that one event caused the other. Scientists must design a controlled investigation to recognize the exact cause and effect.

Draw Conclusions

When scientists have analyzed the data they collected, they proceed to draw conclusions about the data. These conclusions are sometimes stated in words similar to the hypothesis that you formed earlier. They may confirm a hypothesis, or lead you to a new hypothesis.

Infer Scientists often make inferences based on their observations. An inference is an attempt to explain observations or to indicate a cause. An inference is not a fact, but a logical conclusion that needs further investigation. For example, you may infer that a fire has caused smoke. Until you investigate, however, you do not know for sure.

Apply When you draw a conclusion, you must apply those conclusions to determine whether the data supports the hypothesis. If your data do not support your hypothesis, it does not mean that the hypothesis is wrong. It means only that the result of the investigation did not support the hypothesis. Maybe the experiment needs to be redesigned, or some of the initial observations on which the hypothesis was based were incomplete or biased. Perhaps more observation or research is needed to refine your hypothesis. A successful investigation does not always come out the way you originally predicted.

Avoid Bias Sometimes a scientific investigation involves making judgments. When you make a judgment, you form an opinion. It is important to be honest and not to allow any expectations of results to bias your judgments. This is important throughout the entire investigation, from researching to collecting data to drawing conclusions.

Communicate

The communication of ideas is an important part of the work of scientists. A discovery that is not reported will not advance the scientific community's understanding or knowledge. Communication among scientists also is important as a way of improving their investigations.

Scientists communicate in many ways, from writing articles in journals and magazines that explain their investigations and experiments, to announcing important discoveries on television and radio. Scientists also share ideas with colleagues on the Internet or present them as lectures, like the student is doing in **Figure 15.**

Figure 15 A student communicates to his peers about his investigation.

SAFETY SYMBOLS

SAFETY SYMBOLS	HAZARD	EXAMPLES	PRECAUTION	REMEDY
DISPOSAL	Special disposal procedures need to be followed.	certain chemicals, living organisms	Do not dispose of these materials in the sink or trash can.	Dispose of wastes as directed by your teacher.
BIOLOGICAL	Organisms or other biological materials that might be harmful to humans	bacteria, fungi, blood, unpreserved tissues, plant materials	Avoid skin contact with these materials. Wear mask or gloves.	Notify your teacher if you suspect contact with material. Wash hands thoroughly.
EXTREME TEMPERATURE	Objects that can burn skin by being too cold or too hot	boiling liquids, hot plates, dry ice, liquid nitrogen	Use proper protection when handling.	Go to your teacher for first aid.
SHARP OBJECT	Use of tools or glassware that can easily puncture or slice skin	razor blades, pins, scalpels, pointed tools, dissecting probes, broken glass	Practice common-sense behavior and follow guidelines for use of the tool.	Go to your teacher for first aid.
FUME	Possible danger to respiratory tract from fumes	ammonia, acetone, nail polish remover, heated sulfur, moth balls	Make sure there is good ventilation. Never smell fumes directly. Wear a mask.	Leave foul area and notify your teacher immediately.
ELECTRICAL	Possible danger from electrical shock or burn	improper grounding, liquid spills, short circuits, exposed wires	Double-check setup with teacher. Check condition of wires and apparatus.	Do not attempt to fix electrical problems. Notify your teacher immediately.
IRRITANT	Substances that can irritate the skin or mucous membranes of the respiratory tract	pollen, moth balls, steel wool, fiberglass, potassium permanganate	Wear dust mask and gloves. Practice extra care when handling these materials.	Go to your teacher for first aid.
CHEMICAL	Chemicals can react with and destroy tissue and other materials	bleaches such as hydrogen peroxide; acids such as sulfuric acid, hydrochloric acid; bases such as ammonia, sodium hydroxide	Wear goggles, gloves, and an apron.	Immediately flush the affected area with water and notify your teacher.
TOXIC	Substance may be poisonous if touched, inhaled, or swallowed.	mercury, many metal compounds, iodine, poinsettia plant parts	Follow your teacher's instructions.	Always wash hands thoroughly after use. Go to your teacher for first aid.
FLAMMABLE	Flammable chemicals may be ignited by open flame, spark, or exposed heat.	alcohol, kerosene, potassium permanganate	Avoid open flames and heat when using flammable chemicals.	Notify your teacher immediately. Use fire safety equipment if applicable.
OPEN FLAME	Open flame in use, may cause fire.	hair, clothing, paper, synthetic materials	Tie back hair and loose clothing. Follow teacher's instruction on lighting and extinguishing flames.	Notify your teacher immediately. Use fire safety equipment if applicable.

 Eye Safety Proper eye protection should be worn at all times by anyone performing or observing science activities.

 Clothing Protection This symbol appears when substances could stain or burn clothing.

 Animal Safety This symbol appears when safety of animals and students must be ensured.

 Handwashing After the lab, wash hands with soap and water before removing goggles.

Safety in the Science Laboratory

The science laboratory is a safe place to work if you follow standard safety procedures. Being responsible for your own safety helps to make the entire laboratory a safer place for everyone. When performing any lab, read and apply the caution statements and safety symbol listed at the beginning of the lab.

General Safety Rules

1. Obtain your teacher's permission to begin all investigations and use laboratory equipment.

2. Study the procedure. Ask your teacher any questions. Be sure you understand safety symbols shown on the page.

3. Notify your teacher about allergies or other health conditions which can affect your participation in a lab.

4. Learn and follow use and safety procedures for your equipment. If unsure, ask your teacher.

5. Never eat, drink, chew gum, apply cosmetics, or do any personal grooming in the lab. Never use lab glassware as food or drink containers. Keep your hands away from your face and mouth.

6. Know the location and proper use of the safety shower, eye wash, fire blanket, and fire alarm.

Prevent Accidents

1. Use the safety equipment provided to you. Goggles and a safety apron should be worn during investigations.

2. Do NOT use hair spray, mousse, or other flammable hair products. Tie back long hair and tie down loose clothing.

3. Do NOT wear sandals or other open-toed shoes in the lab.

4. Remove jewelry on hands and wrists. Loose jewelry, such as chains and long necklaces, should be removed to prevent them from getting caught in equipment.

5. Do not taste any substances or draw any material into a tube with your mouth.

6. Proper behavior is expected in the lab. Practical jokes and fooling around can lead to accidents and injury.

7. Keep your work area uncluttered.

Laboratory Work

1. Collect and carry all equipment and materials to your work area before beginning a lab.

2. Remain in your own work area unless given permission by your teacher to leave it.

3. Dispose of chemicals and other materials as directed by your teacher. Place broken glass and solid substances in the proper containers. Never discard materials in the sink.

4. Clean your work area.

5. Wash your hands with soap and water thoroughly BEFORE removing your goggles.

Emergencies

1. Report any fire, electrical shock, glassware breakage, spill, or injury, no matter how small, to your teacher immediately. Follow his or her instructions.

2. If your clothing should catch fire, STOP, DROP, and ROLL. If possible, smother it with the fire blanket or get under a safety shower. NEVER RUN.

3. If a fire should occur, turn off all gas and leave the room according to established procedures.

4. In most instances, your teacher will clean up spills. Do NOT attempt to clean up spills unless you are given permission and instructions to do so.

5. If chemicals come into contact with your eyes or skin, notify your teacher immediately. Use the eyewash or flush your skin or eyes with large quantities of water.

6. The fire extinguisher and first-aid kit should only be used by your teacher unless it is an extreme emergency and you have been given permission.

7. If someone is injured or becomes ill, only a professional medical provider or someone certified in first aid should perform first-aid procedures.

3. Always slant test tubes away from yourself and others when heating them, adding substances to them, or rinsing them.

4. If instructed to smell a substance in a container, hold the container a short distance away and fan vapors towards your nose.

5. Do NOT substitute other chemicals/substances for those in the materials list unless instructed to do so by your teacher.

6. Do NOT take any materials or chemicals outside of the laboratory.

7. Stay out of storage areas unless instructed to be there and supervised by your teacher.

Laboratory Cleanup

1. Turn off all burners, water, and gas, and disconnect all electrical devices.

2. Clean all pieces of equipment and return all materials to their proper places.

EXTRA  Labs

From Your Kitchen, Junk Drawer, or Yard

1 Bending Water

▶ **Real-World Question**

How can a plastic rod bend water without touching it?

Possible Materials 🥽 ✋

- plastic rod
- plastic clothes hanger
- 100% wool clothing
- water faucet

▶ **Procedure**

1. Turn on a faucet until a narrow, smooth stream of water is flowing out of it. The stream of water cannot be too wide, and it cannot flow in a broken pattern.

2. Vigorously rub a plastic rod on a piece of 100% wool clothing for about 15 s.

3. Immediately hold the rod near the center of the stream of water. Move the rod close to the stream. Do not touch the water.

4. Observe what happens to the water.

▶ **Conclude and Apply**

1. Describe how the plastic rod affected the stream of water.

2. Explain why the plastic rod affected the water.

2 Testing Magnets

▶ **Real-World Question**

How do the strengths of kitchen magnets compare?

Possible Materials 🧲 🥽 ✋

- several kitchen magnets
- metric ruler
- small pin or paper clip

▶ **Procedure**

1. Place a small pin or paper clip on a flat, nonmetallic surface such as a wooden table.

2. Holding your metric ruler vertically, place it next to the pin with the 0 cm mark on the tabletop.

3. Hold a kitchen magnet at the 10 cm mark on the ruler.

4. Slowly lower the magnet toward the pin. At the point where the pin is attracted to the magnet, measure the height of the magnet from the table. Record the height in your Science Journal.

5. Repeat steps 2–4 to test your other kitchen magnets.

▶ **Conclude and Apply**

1. Describe the results of your experiment.

2. Infer how the kitchen magnets should be used based on the results of your experiment.

Adult supervision required for all labs.

3 Pattern Counting

▶ *Real-World Question*

What pattern is used to count in binary?

▶ *Procedure*

1. Study the pattern used for counting in 4-bit binary.
2. Describe the pattern in your own words.
3. To test your understanding of the pattern, close the book and write the counting pattern from 0–15 by using your notes.

▶ *Conclude and Apply*

1. Can this pattern continue past 15? Explain.
2. Develop a pattern to count to 32.

Decimal Number	Binary Number	Decimal Number	Binary Number
0	0000	8	1000
1	0001	9	1001
2	0010	10	1010
3	0011	11	1011
4	0100	12	1100
5	0101	13	1101
6	0110	14	1110
7	0111	15	1111

Computer Skills

People who study science rely on computers, like the one in **Figure 16,** to record and store data and to analyze results from investigations. Whether you work in a laboratory or just need to write a lab report with tables, good computer skills are a necessity.

Using the computer comes with responsibility. Issues of ownership, security, and privacy can arise. Remember, if you did not author the information you are using, you must provide a source for your information. Also, anything on a computer can be accessed by others. Do not put anything on the computer that you would not want everyone to know. To add more security to your work, use a password.

Use a Word Processing Program

A computer program that allows you to type your information, change it as many times as you need to, and then print it out is called a word processing program. Word processing programs also can be used to make tables.

Figure 16 A computer will make reports neater and more professional looking.

Learn the Skill To start your word processing program, a blank document, sometimes called "Document 1," appears on the screen. To begin, start typing. To create a new document, click the *New* button on the standard tool bar. These tips will help you format the document.

- The program will automatically move to the next line; press *Enter* if you wish to start a new paragraph.
- Symbols, called non-printing characters, can be hidden by clicking the *Show/Hide* button on your toolbar.
- To insert text, move the cursor to the point where you want the insertion to go, click on the mouse once, and type the text.
- To move several lines of text, select the text and click the *Cut* button on your toolbar. Then position your cursor in the location that you want to move the cut text and click *Paste.* If you move to the wrong place, click *Undo.*
- The spell check feature does not catch words that are misspelled to look like other words, like "cold" instead of "gold." Always reread your document to catch all spelling mistakes.
- To learn about other word processing methods, read the user's manual or click on the *Help* button.
- You can integrate databases, graphics, and spreadsheets into documents by copying from another program and pasting it into your document, or by using desktop publishing (DTP). DTP software allows you to put text and graphics together to finish your document with a professional look. This software varies in how it is used and its capabilities.

Use a Database

A collection of facts stored in a computer and sorted into different fields is called a database. A database can be reorganized in any way that suits your needs.

Learn the Skill A computer program that allows you to create your own database is a database management system (DBMS). It allows you to add, delete, or change information. Take time to get to know the features of your database software.

- Determine what facts you would like to include and research to collect your information.
- Determine how you want to organize the information.
- Follow the instructions for your particular DBMS to set up fields. Then enter each item of data in the appropriate field.
- Follow the instructions to sort the information in order of importance.
- Evaluate the information in your database, and add, delete, or change as necessary.

Use the Internet

The Internet is a global network of computers where information is stored and shared. To use the Internet, like the students in **Figure 17,** you need a modem to connect your computer to a phone line and an Internet Service Provider account.

Learn the Skill To access internet sites and information, use a "Web browser," which lets you view and explore pages on the World Wide Web. Each page is its own site, and each site has its own address, called a URL. Once you have found a Web browser, follow these steps for a search (this also is how you search a database).

Figure 17 The Internet allows you to search a global network for a variety of information.

- Be as specific as possible. If you know you want to research "gold," don't type in "elements." Keep narrowing your search until you find what you want.
- Web sites that end in *.com* are commercial Web sites; *.org, .edu,* and *.gov* are nonprofit, educational, or government Web sites.
- Electronic encyclopedias, almanacs, indexes, and catalogs will help locate and select relevant information.
- Develop a "home page" with relative ease. When developing a Web site, NEVER post pictures or disclose personal information such as location, names, or phone numbers. Your school or community usually can host your Web site. A basic understanding of HTML (hypertext mark-up language), the language of Web sites, is necessary. Software that creates HTML code is called authoring software, and can be downloaded free from many Web sites. This software allows text and pictures to be arranged as the software is writing the HTML code.

Use a Spreadsheet

A spreadsheet, shown in **Figure 18,** can perform mathematical functions with any data arranged in columns and rows. By entering a simple equation into a cell, the program can perform operations in specific cells, rows, or columns.

Learn the Skill Each column (vertical) is assigned a letter, and each row (horizontal) is assigned a number. Each point where a row and column intersect is called a cell, and is labeled according to where it is located—Column A, Row 1 (A1).

■ Decide how to organize the data, and enter it in the correct row or column.

■ Spreadsheets can use standard formulas or formulas can be customized to calculate cells.

■ To make a change, click on a cell to make it activate, and enter the edited data or formula.

■ Spreadsheets also can display your results in graphs. Choose the style of graph that best represents the data.

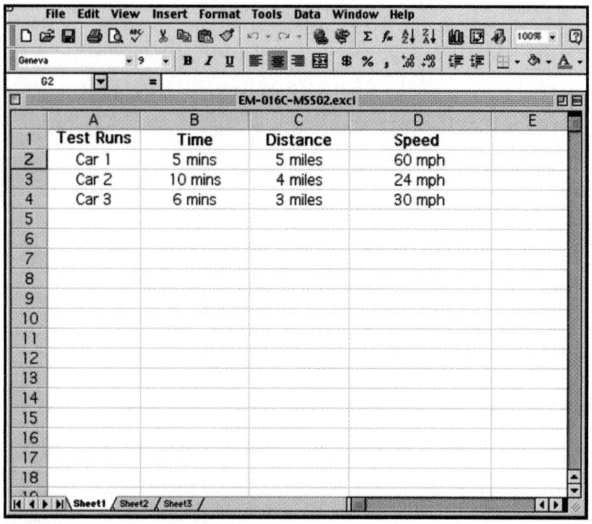

Figure 18 A spreadsheet allows you to perform mathematical operations on your data.

Use Graphics Software

Adding pictures, called graphics, to your documents is one way to make your documents more meaningful and exciting. This software adds, edits, and even constructs graphics. There is a variety of graphics software programs. The tools used for drawing can be a mouse, keyboard, or other specialized devices. Some graphics programs are simple. Others are complicated, called computer-aided design (CAD) software.

Learn the Skill It is important to have an understanding of the graphics software being used before starting. The better the software is understood, the better the results. The graphics can be placed in a word-processing document.

■ Clip art can be found on a variety of internet sites, and on CDs. These images can be copied and pasted into your document.

■ When beginning, try editing existing drawings, then work up to creating drawings.

■ The images are made of tiny rectangles of color called pixels. Each pixel can be altered.

■ Digital photography is another way to add images. The photographs in the memory of a digital camera can be downloaded into a computer, then edited and added to the document.

■ Graphics software also can allow animation. The software allows drawings to have the appearance of movement by connecting basic drawings automatically. This is called in-betweening, or tweening.

■ Remember to save often.

Presentation Skills

Develop Multimedia Presentations

Most presentations are more dynamic if they include diagrams, photographs, videos, or sound recordings, like the one shown in **Figure 19.** A multimedia presentation involves using stereos, overhead projectors, televisions, computers, and more.

Learn the Skill Decide the main points of your presentation, and what types of media would best illustrate those points.

- Make sure you know how to use the equipment you are working with.
- Practice the presentation using the equipment several times.
- Enlist the help of a classmate to push play or turn lights out for you. Be sure to practice your presentation with him or her.
- If possible, set up all of the equipment ahead of time, and make sure everything is working properly.

Figure 19 These students are engaging the audience using a variety of tools.

Computer Presentations

There are many different interactive computer programs that you can use to enhance your presentation. Most computers have a compact disc (CD) drive that can play both CDs and digital video discs (DVDs). Also, there is hardware to connect a regular CD, DVD, or VCR. These tools will enhance your presentation.

Another method of using the computer to aid in your presentation is to develop a slide show using a computer program. This can allow movement of visuals at the presenter's pace, and can allow for visuals to build on one another.

Learn the Skill In order to create multimedia presentations on a computer, you need to have certain tools. These may include traditional graphic tools and drawing programs, animation programs, and authoring systems that tie everything together. Your computer will tell you which tools it supports. The most important step is to learn about the tools that you will be using.

- Often, color and strong images will convey a point better than words alone. Use the best methods available to convey your point.
- As with other presentations, practice many times.
- Practice your presentation with the tools you and any assistants will be using.
- Maintain eye contact with the audience. The purpose of using the computer is not to prompt the presenter, but to help the audience understand the points of the presentation.

Math Review

Use Fractions

A fraction compares a part to a whole. In the fraction $\frac{2}{3}$, the 2 represents the part and is the numerator. The 3 represents the whole and is the denominator.

Reduce Fractions To reduce a fraction, you must find the largest factor that is common to both the numerator and the denominator, the greatest common factor (GCF). Divide both numbers by the GCF. The fraction has then been reduced, or it is in its simplest form.

Example Twelve of the 20 chemicals in the science lab are in powder form. What fraction of the chemicals used in the lab are in powder form?

Step 1 Write the fraction.

$$\frac{part}{whole} = \frac{12}{20}$$

Step 2 To find the GCF of the numerator and denominator, list all of the factors of each number.

Factors of 12: 1, 2, 3, 4, 6, 12 (the numbers that divide evenly into 12)

Factors of 20: 1, 2, 4, 5, 10, 20 (the numbers that divide evenly into 20)

Step 3 List the common factors.

1, 2, 4.

Step 4 Choose the greatest factor in the list.

The GCF of 12 and 20 is 4.

Step 5 Divide the numerator and denominator by the GCF.

$$\frac{12 \div 4}{20 \div 4} = \frac{3}{5}$$

In the lab, $\frac{3}{5}$ of the chemicals are in powder form.

Practice Problem At an amusement park, 66 of 90 rides have a height restriction. What fraction of the rides, in its simplest form, has a height restriction?

Add and Subtract Fractions To add or subtract fractions with the same denominator, add or subtract the numerators and write the sum or difference over the denominator. After finding the sum or difference, find the simplest form for your fraction.

Example 1 In the forest outside your house, $\frac{1}{8}$ of the animals are rabbits, $\frac{3}{8}$ are squirrels, and the remainder are birds and insects. How many are mammals?

Step 1 Add the numerators.

$$\frac{1}{8} + \frac{3}{8} = \frac{(1 + 3)}{8} = \frac{4}{8}$$

Step 2 Find the GCF.

$$\frac{4}{8} \text{ (GCF, 4)}$$

Step 3 Divide the numerator and denominator by the GCF.

$$\frac{4}{4} = 1, \ \frac{8}{4} = 2$$

$\frac{1}{2}$ of the animals are mammals.

Example 2 If $\frac{7}{16}$ of the Earth is covered by freshwater, and $\frac{1}{16}$ of that is in glaciers, how much freshwater is not frozen?

Step 1 Subtract the numerators.

$$\frac{7}{16} - \frac{1}{16} = \frac{(7 - 1)}{16} = \frac{6}{16}$$

Step 2 Find the GCF.

$$\frac{6}{16} \text{ (GCF, 2)}$$

Step 3 Divide the numerator and denominator by the GCF.

$$\frac{6}{2} = 3, \ \frac{16}{2} = 8$$

$\frac{3}{8}$ of the freshwater is not frozen.

Practice Problem A bicycle rider is going 15 km/h for $\frac{4}{9}$ of his ride, 10 km/h for $\frac{2}{9}$ of his ride, and 8 km/h for the remainder of the ride. How much of his ride is he going over 8 km/h?

Unlike Denominators To add or subtract fractions with unlike denominators, first find the least common denominator (LCD). This is the smallest number that is a common multiple of both denominators. Rename each fraction with the LCD, and then add or subtract. Find the simplest form if necessary.

Example 1 A chemist makes a paste that is $\frac{1}{2}$ table salt (NaCl), $\frac{1}{3}$ sugar ($C_6H_{12}O_6$), and the rest water (H_2O). How much of the paste is a solid?

Step 1 Find the LCD of the fractions.

$\frac{1}{2} + \frac{1}{3}$ (LCD, 6)

Step 2 Rename each numerator and each denominator with the LCD.

$1 \times 3 = 3, \ 2 \times 3 = 6$

$1 \times 2 = 2, \ 3 \times 2 = 6$

Step 3 Add the numerators.

$\frac{3}{6} + \frac{2}{6} = \frac{(3 + 2)}{6} = \frac{5}{6}$

$\frac{5}{6}$ of the paste is a solid.

Example 2 The average precipitation in Grand Junction, CO, is $\frac{7}{10}$ inch in November, and $\frac{3}{5}$ inch in December. What is the total average precipitation?

Step 1 Find the LCD of the fractions.

$\frac{7}{10} + \frac{3}{5}$ (LCD, 10)

Step 2 Rename each numerator and each denominator with the LCD.

$7 \times 1 = 7, \ 10 \times 1 = 10$

$3 \times 2 = 6, \ 5 \times 2 = 10$

Step 3 Add the numerators.

$\frac{7}{10} + \frac{6}{10} = \frac{(7 + 6)}{10} = \frac{13}{10}$

$\frac{13}{10}$ inches total precipitation, or $1\frac{3}{10}$ inches.

Practice Problem On an electric bill, about $\frac{1}{8}$ of the energy is from solar energy and about $\frac{1}{10}$ is from wind power. How much of the total bill is from solar energy and wind power combined?

Example 3 In your body, $\frac{7}{10}$ of your muscle contractions are involuntary (cardiac and smooth muscle tissue). Smooth muscle makes $\frac{3}{15}$ of your muscle contractions. How many of your muscle contractions are made by cardiac muscle?

Step 1 Find the LCD of the fractions.

$\frac{7}{10} - \frac{3}{15}$ (LCD, 30)

Step 2 Rename each numerator and each denominator with the LCD.

$7 \times 3 = 21, \ 10 \times 3 = 30$

$3 \times 2 = 6, \ 15 \times 2 = 30$

Step 3 Subtract the numerators.

$\frac{21}{30} - \frac{6}{30} = \frac{(21 - 6)}{30} = \frac{15}{30}$

Step 4 Find the GCF.

$\frac{15}{30}$ (GCF, 15)

$\frac{1}{2}$

$\frac{1}{2}$ of all muscle contractions are cardiac muscle.

Example 4 Tony wants to make cookies that call for $\frac{3}{4}$ of a cup of flour, but he only has $\frac{1}{3}$ of a cup. How much more flour does he need?

Step 1 Find the LCD of the fractions.

$\frac{3}{4} - \frac{1}{3}$ (LCD, 12)

Step 2 Rename each numerator and each denominator with the LCD.

$3 \times 3 = 9, \ 4 \times 3 = 12$

$1 \times 4 = 4, \ 3 \times 4 = 12$

Step 3 Subtract the numerators.

$\frac{9}{12} - \frac{4}{12} = \frac{(9 - 4)}{12} = \frac{5}{12}$

$\frac{5}{12}$ of a cup of flour.

Practice Problem Using the information provided to you in Example 3 above, determine how many muscle contractions are voluntary (skeletal muscle).

Multiply Fractions To multiply with fractions, multiply the numerators and multiply the denominators. Find the simplest form if necessary.

Example Multiply $\frac{3}{5}$ by $\frac{1}{3}$.

Step 1 Multiply the numerators and denominators.

$$\frac{3}{5} \times \frac{1}{3} = \frac{(3 \times 1)}{(5 \times 3)} = \frac{3}{15}$$

Step 2 Find the GCF.

$$\frac{3}{15} \quad (GCF, 3)$$

Step 3 Divide the numerator and denominator by the GCF.

$$\frac{3}{3} = 1, \quad \frac{15}{3} = 5$$

$$\frac{1}{5}$$

$\frac{3}{5}$ multiplied by $\frac{1}{3}$ is $\frac{1}{5}$.

Practice Problem Multiply $\frac{3}{14}$ by $\frac{5}{16}$.

Find a Reciprocal Two numbers whose product is 1 are called multiplicative inverses, or reciprocals.

Example Find the reciprocal of $\frac{3}{8}$.

Step 1 Inverse the fraction by putting the denominator on top and the numerator on the bottom.

$$\frac{8}{3}$$

The reciprocal of $\frac{3}{8}$ is $\frac{8}{3}$.

Practice Problem Find the reciprocal of $\frac{4}{9}$.

Divide Fractions To divide one fraction by another fraction, multiply the dividend by the reciprocal of the divisor. Find the simplest form if necessary.

Example 1 Divide $\frac{1}{9}$ by $\frac{1}{3}$.

Step 1 Find the reciprocal of the divisor.

The reciprocal of $\frac{1}{3}$ is $\frac{3}{1}$.

Step 2 Multiply the dividend by the reciprocal of the divisor.

$$\frac{\frac{1}{9}}{\frac{1}{3}} = \frac{1}{9} \times \frac{3}{1} = \frac{(1 \times 3)}{(9 \times 1)} = \frac{3}{9}$$

Step 3 Find the GCF.

$$\frac{3}{9} \quad (GCF, 3)$$

Step 4 Divide the numerator and denominator by the GCF.

$$\frac{3}{3} = 1, \quad \frac{9}{3} = 3$$

$$\frac{1}{3}$$

$\frac{1}{9}$ divided by $\frac{1}{3}$ is $\frac{1}{3}$.

Example 2 Divide $\frac{3}{5}$ by $\frac{1}{4}$.

Step 1 Find the reciprocal of the divisor.

The reciprocal of $\frac{1}{4}$ is $\frac{4}{1}$.

Step 2 Multiply the dividend by the reciprocal of the divisor.

$$\frac{\frac{3}{5}}{\frac{1}{4}} = \frac{3}{5} \times \frac{4}{1} = \frac{(3 \times 4)}{(5 \times 1)} = \frac{12}{5}$$

$\frac{3}{5}$ divided by $\frac{1}{4}$ is $\frac{12}{5}$ or $2\frac{2}{5}$.

Practice Problem Divide $\frac{3}{11}$ by $\frac{7}{10}$.

Use Ratios

When you compare two numbers by division, you are using a ratio. Ratios can be written 3 to 5, 3:5, or $\frac{3}{5}$. Ratios, like fractions, also can be written in simplest form.

Ratios can represent probabilities, also called odds. This is a ratio that compares the number of ways a certain outcome occurs to the number of outcomes. For example, if you flip a coin 100 times, what are the odds that it will come up heads? There are two possible outcomes, heads or tails, so the odds of coming up heads are 50:100. Another way to say this is that 50 out of 100 times the coin will come up heads. In its simplest form, the ratio is 1:2.

Example 1 A chemical solution contains 40 g of salt and 64 g of baking soda. What is the ratio of salt to baking soda as a fraction in simplest form?

Step 1 Write the ratio as a fraction.
$$\frac{\text{salt}}{\text{baking soda}} = \frac{40}{64}$$

Step 2 Express the fraction in simplest form.
The GCF of 40 and 64 is 8.
$$\frac{40}{64} = \frac{40 \div 8}{64 \div 8} = \frac{5}{8}$$

The ratio of salt to baking soda in the sample is 5:8.

Example 2 Sean rolls a 6-sided die 6 times. What are the odds that the side with a 3 will show?

Step 1 Write the ratio as a fraction.
$$\frac{\text{number of sides with a 3}}{\text{number of sides}} = \frac{1}{6}$$

Step 2 Multiply by the number of attempts.
$$\frac{1}{6} \times 6 \text{ attempts} = \frac{6}{6} \text{ attempts} = 1 \text{ attempt}$$

1 attempt out of 6 will show a 3.

Practice Problem Two metal rods measure 100 cm and 144 cm in length. What is the ratio of their lengths in simplest form?

Use Decimals

A fraction with a denominator that is a power of ten can be written as a decimal. For example, 0.27 means $\frac{27}{100}$. The decimal point separates the ones place from the tenths place.

Any fraction can be written as a decimal using division. For example, the fraction $\frac{5}{8}$ can be written as a decimal by dividing 5 by 8. Written as a decimal, it is 0.625.

Add or Subtract Decimals When adding and subtracting decimals, line up the decimal points before carrying out the operation.

Example 1 Find the sum of 47.68 and 7.80.

Step 1 Line up the decimal places when you write the numbers.

$$\begin{array}{r} 47.68 \\ + \ 7.80 \\ \hline \end{array}$$

Step 2 Add the decimals.

$$\begin{array}{r} 47.68 \\ + \ 7.80 \\ \hline 55.48 \end{array}$$

The sum of 47.68 and 7.80 is 55.48.

Example 2 Find the difference of 42.17 and 15.85.

Step 1 Line up the decimal places when you write the number.

$$\begin{array}{r} 42.17 \\ -15.85 \\ \hline \end{array}$$

Step 2 Subtract the decimals.

$$\begin{array}{r} 42.17 \\ -15.85 \\ \hline 26.32 \end{array}$$

The difference of 42.17 and 15.85 is 26.32.

Practice Problem Find the sum of 1.245 and 3.842.

Multiply Decimals To multiply decimals, multiply the numbers like any other number, ignoring the decimal point. Count the decimal places in each factor. The product will have the same number of decimal places as the sum of the decimal places in the factors.

Example Multiply 2.4 by 5.9.

Step 1 Multiply the factors like two whole numbers.
$24 \times 59 = 1416$

Step 2 Find the sum of the number of decimal places in the factors. Each factor has one decimal place, for a sum of two decimal places.

Step 3 The product will have two decimal places.
14.16

The product of 2.4 and 5.9 is 14.16.

Practice Problem Multiply 4.6 by 2.2.

Divide Decimals When dividing decimals, change the divisor to a whole number. To do this, multiply both the divisor and the dividend by the same power of ten. Then place the decimal point in the quotient directly above the decimal point in the dividend. Then divide as you do with whole numbers.

Example Divide 8.84 by 3.4.

Step 1 Multiply both factors by 10.
$3.4 \times 10 = 34, 8.84 \times 10 = 88.4$

Step 2 Divide 88.4 by 34.

$$
\begin{array}{r}
2.6 \\
34\overline{)88.4} \\
-68 \\
\hline
204 \\
-204 \\
\hline
0
\end{array}
$$

8.84 divided by 3.4 is 2.6.

Practice Problem Divide 75.6 by 3.6.

Use Proportions

An equation that shows that two ratios are equivalent is a proportion. The ratios $\frac{2}{4}$ and $\frac{5}{10}$ are equivalent, so they can be written as $\frac{2}{4} = \frac{5}{10}$. This equation is a proportion.

When two ratios form a proportion, the cross products are equal. To find the cross products in the proportion $\frac{2}{4} = \frac{5}{10}$, multiply the 2 and the 10, and the 4 and the 5. Therefore $2 \times 10 = 4 \times 5$, or $20 = 20$.

Because you know that both proportions are equal, you can use cross products to find a missing term in a proportion. This is known as solving the proportion.

Example The heights of a tree and a pole are proportional to the lengths of their shadows. The tree casts a shadow of 24 m when a 6-m pole casts a shadow of 4 m. What is the height of the tree?

Step 1 Write a proportion.
$$\frac{\text{height of tree}}{\text{height of pole}} = \frac{\text{length of tree's shadow}}{\text{length of pole's shadow}}$$

Step 2 Substitute the known values into the proportion. Let h represent the unknown value, the height of the tree.
$$\frac{h}{6} = \frac{24}{4}$$

Step 3 Find the cross products.
$$h \times 4 = 6 \times 24$$

Step 4 Simplify the equation.
$$4h = 144$$

Step 5 Divide each side by 4.
$$\frac{4h}{4} = \frac{144}{4}$$
$$h = 36$$

The height of the tree is 36 m.

Practice Problem The ratios of the weights of two objects on the Moon and on Earth are in proportion. A rock weighing 3 N on the Moon weighs 18 N on Earth. How much would a rock that weighs 5 N on the Moon weigh on Earth?

Use Percentages

The word *percent* means "out of one hundred." It is a ratio that compares a number to 100. Suppose you read that 77 percent of the Earth's surface is covered by water. That is the same as reading that the fraction of the Earth's surface covered by water is $\frac{77}{100}$. To express a fraction as a percent, first find the equivalent decimal for the fraction. Then, multiply the decimal by 100 and add the percent symbol.

Example Express $\frac{13}{20}$ as a percent.

Step 1 Find the equivalent decimal for the fraction.

$$
\begin{array}{r}
0.65 \\
20\overline{)13.00} \\
12\,0 \\
\hline
1\,00 \\
1\,00 \\
\hline
0
\end{array}
$$

Step 2 Rewrite the fraction $\frac{13}{20}$ as 0.65.

Step 3 Multiply 0.65 by 100 and add the % sign.

$$0.65 \times 100 = 65 = 65\%$$

So, $\frac{13}{20} = 65\%$.

This also can be solved as a proportion.

Example Express $\frac{13}{20}$ as a percent.

Step 1 Write a proportion.

$$\frac{13}{20} = \frac{x}{100}$$

Step 2 Find the cross products.

$$1300 = 20x$$

Step 3 Divide each side by 20.

$$\frac{1300}{20} = \frac{20x}{20}$$
$$65\% = x$$

Practice Problem In one year, 73 of 365 days were rainy in one city. What percent of the days in that city were rainy?

Solve One-Step Equations

A statement that two things are equal is an equation. For example, $A = B$ is an equation that states that A is equal to B.

An equation is solved when a variable is replaced with a value that makes both sides of the equation equal. To make both sides equal the inverse operation is used. Addition and subtraction are inverses, and multiplication and division are inverses.

Example 1 Solve the equation $x - 10 = 35$.

Step 1 Find the solution by adding 10 to each side of the equation.

$$x - 10 = 35$$
$$x - 10 + 10 = 35 + 10$$
$$x = 45$$

Step 2 Check the solution.

$$x - 10 = 35$$
$$45 - 10 = 35$$
$$35 = 35$$

Both sides of the equation are equal, so $x = 45$.

Example 2 In the formula $a = bc$, find the value of c if $a = 20$ and $b = 2$.

Step 1 Rearrange the formula so the unknown value is by itself on one side of the equation by dividing both sides by b.

$$a = bc$$
$$\frac{a}{b} = \frac{bc}{b}$$
$$\frac{a}{b} = c$$

Step 2 Replace the variables a and b with the values that are given.

$$\frac{a}{b} = c$$
$$\frac{20}{2} = c$$
$$10 = c$$

Step 3 Check the solution.

$$a = bc$$
$$20 = 2 \times 10$$
$$20 = 20$$

Both sides of the equation are equal, so $c = 10$ is the solution when $a = 20$ and $b = 2$.

Practice Problem In the formula $h = gd$, find the value of d if $g = 12.3$ and $h = 17.4$.

Use Statistics

The branch of mathematics that deals with collecting, analyzing, and presenting data is statistics. In statistics, there are three common ways to summarize data with a single number—the mean, the median, and the mode.

The **mean** of a set of data is the arithmetic average. It is found by adding the numbers in the data set and dividing by the number of items in the set.

The **median** is the middle number in a set of data when the data are arranged in numerical order. If there were an even number of data points, the median would be the mean of the two middle numbers.

The **mode** of a set of data is the number or item that appears most often.

Another number that often is used to describe a set of data is the range. The **range** is the difference between the largest number and the smallest number in a set of data.

A **frequency table** shows how many times each piece of data occurs, usually in a survey. **Table 2** below shows the results of a student survey on favorite color.

Table 2 Student Color Choice		
Color	**Tally**	**Frequency**
red	||||	4
blue	卌	5
black	||	2
green	|||	3
purple	卌 ||	7
yellow	卌 |	6

Based on the frequency table data, which color is the favorite?

Example The speeds (in m/s) for a race car during five different time trials are 39, 37, 44, 36, and 44.

To find the mean:

Step 1 Find the sum of the numbers.

$$39 + 37 + 44 + 36 + 44 = 200$$

Step 2 Divide the sum by the number of items, which is 5.

$$200 \div 5 = 40$$

The mean is 40 m/s.

To find the median:

Step 1 Arrange the measures from least to greatest.

36, 37, 39, 44, 44

Step 2 Determine the middle measure.

36, 37, <u>39</u>, 44, 44

The median is 39 m/s.

To find the mode:

Step 1 Group the numbers that are the same together.

44, 44, 36, 37, 39

Step 2 Determine the number that occurs most in the set.

<u>44, 44</u>, 36, 37, 39

The mode is 44 m/s.

To find the range:

Step 1 Arrange the measures from largest to smallest.

44, 44, 39, 37, 36

Step 2 Determine the largest and smallest measures in the set.

<u>44</u>, 44, 39, 37, <u>36</u>

Step 3 Find the difference between the largest and smallest measures.

$$44 - 36 = 8$$

The range is 8 m/s.

Practice Problem Find the mean, median, mode, and range for the data set 8, 4, 12, 8, 11, 14, 16.

Use Geometry

The branch of mathematics that deals with the measurement, properties, and relationships of points, lines, angles, surfaces, and solids is called geometry.

Perimeter The **perimeter** (P) is the distance around a geometric figure. To find the perimeter of a rectangle, add the length and width and multiply that sum by two, or $2(l + w)$. To find perimeters of irregular figures, add the length of the sides.

Example 1 Find the perimeter of a rectangle that is 3 m long and 5 m wide.

Step 1 You know that the perimeter is 2 times the sum of the width and length.
$$P = 2(3\text{ m} + 5\text{ m})$$

Step 2 Find the sum of the width and length.
$$P = 2(8\text{ m})$$

Step 3 Multiply by 2.
$$P = 16\text{ m}$$

The perimeter is 16 m.

Example 2 Find the perimeter of a shape with sides measuring 2 cm, 5 cm, 6 cm, 3 cm.

Step 1 You know that the perimeter is the sum of all the sides.
$$P = 2 + 5 + 6 + 3$$

Step 2 Find the sum of the sides.
$$P = 2 + 5 + 6 + 3$$
$$P = 16$$

The perimeter is 16 cm.

Practice Problem Find the perimeter of a rectangle with a length of 18 m and a width of 7 m.

Practice Problem Find the perimeter of a triangle measuring 1.6 cm by 2.4 cm by 2.4 cm.

Area of a Rectangle The **area** (A) is the number of square units needed to cover a surface. To find the area of a rectangle, multiply the length times the width, or $l \times w$. When finding area, the units also are multiplied. Area is given in square units.

Example Find the area of a rectangle with a length of 1 cm and a width of 10 cm.

Step 1 You know that the area is the length multiplied by the width.
$$A = (1\text{ cm} \times 10\text{ cm})$$

Step 2 Multiply the length by the width. Also multiply the units.
$$A = 10\text{ cm}^2$$

The area is 10 cm^2.

Practice Problem Find the area of a square whose sides measure 4 m.

Area of a Triangle To find the area of a triangle, use the formula:

$$A = \frac{1}{2}(\text{base} \times \text{height})$$

The base of a triangle can be any of its sides. The height is the perpendicular distance from a base to the opposite endpoint, or vertex.

Example Find the area of a triangle with a base of 18 m and a height of 7 m.

Step 1 You know that the area is $\frac{1}{2}$ the base times the height.
$$A = \frac{1}{2}(18\text{ m} \times 7\text{ m})$$

Step 2 Multiply $\frac{1}{2}$ by the product of 18 $\times$ 7. Multiply the units.
$$A = \frac{1}{2}(126\text{ m}^2)$$
$$A = 63\text{ m}^2$$

The area is 63 m^2.

Practice Problem Find the area of a triangle with a base of 27 cm and a height of 17 cm.

Circumference of a Circle The **diameter** (*d*) of a circle is the distance across the circle through its center, and the **radius** (*r*) is the distance from the center to any point on the circle. The radius is half of the diameter. The distance around the circle is called the **circumference** (C). The formula for finding the circumference is:

$$C = 2\pi r \ \ or \ \ C = \pi d$$

The circumference divided by the diameter is always equal to 3.1415926... This nonterminating and nonrepeating number is represented by the Greek letter π (pi). An approximation often used for π is 3.14.

Example 1 Find the circumference of a circle with a radius of 3 m.

Step 1 You know the formula for the circumference is 2 times the radius times π.
$$C = 2\pi(3)$$

Step 2 Multiply 2 times the radius.
$$C = 6\pi$$

Step 3 Multiply by π.
$$C = 19 \ m$$

The circumference is 19 m.

Example 2 Find the circumference of a circle with a diameter of 24.0 cm.

Step 1 You know the formula for the circumference is the diameter times π.
$$C = \pi(24.0)$$

Step 2 Multiply the diameter by π.
$$C = 75.4 \ cm$$

The circumference is 75.4 cm.

Practice Problem Find the circumference of a circle with a radius of 19 cm.

Area of a Circle The formula for the area of a circle is:
$$A = \pi r^2$$

Example 1 Find the area of a circle with a radius of 4.0 cm.

Step 1 $A = \pi(4.0)^2$

Step 2 Find the square of the radius.
$$A = 16\pi$$

Step 3 Multiply the square of the radius by π.
$$A = 50 \ cm^2$$

The area of the circle is 50 cm^2.

Example 2 Find the area of a circle with a radius of 225 m.

Step 1 $A = \pi(225)^2$

Step 2 Find the square of the radius.
$$A = 50625\pi$$

Step 3 Multiply the square of the radius by π.
$$A = 158962.5$$

The area of the circle is 158,962 m^2.

Example 3 Find the area of a circle whose diameter is 20.0 mm.

Step 1 You know the formula for the area of a circle is the square of the radius times π, and that the radius is half of the diameter.
$$A = \pi\left(\frac{20.0}{2}\right)^2$$

Step 2 Find the radius.
$$A = \pi(10.0)^2$$

Step 3 Find the square of the radius.
$$A = 100\pi$$

Step 4 Multiply the square of the radius by π.
$$A = 314 \ mm^2$$

The area is 314 mm^2.

Practice Problem Find the area of a circle with a radius of 16 m.

Volume The measure of space occupied by a solid is the **volume** (V). To find the volume of a rectangular solid multiply the length times width times height, or $V = l \times w \times h$. It is measured in cubic units, such as cubic centimeters (cm^3).

Example Find the volume of a rectangular solid with a length of 2.0 m, a width of 4.0 m, and a height of 3.0 m.

Step 1 You know the formula for volume is the length times the width times the height.
$$V = 2.0 \text{ m} \times 4.0 \text{ m} \times 3.0 \text{ m}$$

Step 2 Multiply the length times the width times the height.
$$V = 24 \text{ m}^3$$

The volume is 24 m^3.

Practice Problem Find the volume of a rectangular solid that is 8 m long, 4 m wide, and 4 m high.

To find the volume of other solids, multiply the area of the base times the height.

Example 1 Find the volume of a solid that has a triangular base with a length of 8.0 m and a height of 7.0 m. The height of the entire solid is 15.0 m.

Step 1 You know that the base is a triangle, and the area of a triangle is $\frac{1}{2}$ the base times the height, and the volume is the area of the base times the height.
$$V = \left[\frac{1}{2} (b \times h) \right] \times 15$$

Step 2 Find the area of the base.
$$V = \left[\frac{1}{2} (8 \times 7) \right] \times 15$$
$$V = \left(\frac{1}{2} \times 56 \right) \times 15$$

Step 3 Multiply the area of the base by the height of the solid.
$$V = 28 \times 15$$
$$V = 420 \text{ m}^3$$

The volume is 420 m^3.

Example 2 Find the volume of a cylinder that has a base with a radius of 12.0 cm, and a height of 21.0 cm.

Step 1 You know that the base is a circle, and the area of a circle is the square of the radius times π, and the volume is the area of the base times the height.
$$V = (\pi r^2) \times 21$$
$$V = (\pi 12^2) \times 21$$

Step 2 Find the area of the base.
$$V = 144\pi \times 21$$
$$V = 452 \times 21$$

Step 3 Multiply the area of the base by the height of the solid.
$$V = 9490 \text{ cm}^3$$

The volume is 9490 cm^3.

Example 3 Find the volume of a cylinder that has a diameter of 15 mm and a height of 4.8 mm.

Step 1 You know that the base is a circle with an area equal to the square of the radius times π. The radius is one-half the diameter. The volume is the area of the base times the height.
$$V = (\pi r^2) \times 4.8$$
$$V = \left[\pi \left(\frac{1}{2} \times 15 \right)^2 \right] \times 4.8$$
$$V = (\pi 7.5^2) \times 4.8$$

Step 2 Find the area of the base.
$$V = 56.25\pi \times 4.8$$
$$V = 176.63 \times 4.8$$

Step 3 Multiply the area of the base by the height of the solid.
$$V = 847.8$$

The volume is 847.8 mm^3.

Practice Problem Find the volume of a cylinder with a diameter of 7 cm in the base and a height of 16 cm.

Science Applications

Measure in SI

The metric system of measurement was developed in 1795. A modern form of the metric system, called the International System (SI), was adopted in 1960 and provides the standard measurements that all scientists around the world can understand.

The SI system is convenient because unit sizes vary by powers of 10. Prefixes are used to name units. Look at **Table 3** for some common SI prefixes and their meanings.

Table 3 Common SI Prefixes			
Prefix	**Symbol**	**Meaning**	
kilo-	k	1,000	thousand
hecto-	h	100	hundred
deka-	da	10	ten
deci-	d	0.1	tenth
centi-	c	0.01	hundredth
milli-	m	0.001	thousandth

Example How many grams equal one kilogram?

Step 1 Find the prefix *kilo* in **Table 3.**

Step 2 Using **Table 3,** determine the meaning of *kilo.* According to the table, it means 1,000. When the prefix *kilo* is added to a unit, it means that there are 1,000 of the units in a "*kilo*unit."

Step 3 Apply the prefix to the units in the question. The units in the question are grams. There are 1,000 grams in a kilogram.

Practice Problem Is a milligram larger or smaller than a gram? How many of the smaller units equal one larger unit? What fraction of the larger unit does one smaller unit represent?

Dimensional Analysis

Convert SI Units In science, quantities such as length, mass, and time sometimes are measured using different units. A process called dimensional analysis can be used to change one unit of measure to another. This process involves multiplying your starting quantity and units by one or more conversion factors. A conversion factor is a ratio equal to one and can be made from any two equal quantities with different units. If 1,000 mL equal 1 L then two ratios can be made.

$$\frac{1,000 \text{ mL}}{1 \text{ L}} = \frac{1 \text{ L}}{1,000 \text{ mL}} = 1$$

One can covert between units in the SI system by using the equivalents in **Table 3** to make conversion factors.

Example 1 How many cm are in 4 m?

Step 1 Write conversion factors for the units given. From **Table 3,** you know that 100 cm = 1 m. The conversion factors are

$$\frac{100 \text{ cm}}{1 \text{ m}} \quad and \quad \frac{1 \text{ m}}{100 \text{ cm}}$$

Step 2 Decide which conversion factor to use. Select the factor that has the units you are converting from (m) in the denominator and the units you are converting to (cm) in the numerator.

$$\frac{100 \text{ cm}}{1 \text{ m}}$$

Step 3 Multiply the starting quantity and units by the conversion factor. Cancel the starting units with the units in the denominator. There are 400 cm in 4 m.

$$4 \text{ m} \times \frac{100 \text{ cm}}{1 \text{ m}} = 400 \text{ cm}$$

Practice Problem How many milligrams are in one kilogram? (Hint: You will need to use two conversion factors from **Table 3.**)

Table 4 Unit System Equivalents

Type of Measurement	Equivalent
Length	1 in = 2.54 cm 1 yd = 0.91 m 1 mi = 1.61 km
Mass and Weight*	1 oz = 28.35 g 1 lb = 0.45 kg 1 ton (short) = 0.91 tonnes (metric tons) 1 lb = 4.45 N
Volume	$1 \text{ in}^3 = 16.39 \text{ cm}^3$ 1 qt = 0.95 L 1 gal = 3.78 L
Area	$1 \text{ in}^2 = 6.45 \text{ cm}^2$ $1 \text{ yd}^2 = 0.83 \text{ m}^2$ $1 \text{ mi}^2 = 2.59 \text{ km}^2$ 1 acre = 0.40 hectares
Temperature	$°C = \dfrac{(°F - 32)}{1.8}$ $K = °C + 273$

*Weight is measured in standard Earth gravity.

Convert Between Unit Systems **Table 4** gives a list of equivalents that can be used to convert between English and SI units.

Example If a meterstick has a length of 100 cm, how long is the meterstick in inches?

Step 1 Write the conversion factors for the units given. From **Table 4,** 1 in = 2.54 cm.

$$\frac{1 \text{ in}}{2.54 \text{ cm}} \quad and \quad \frac{2.54 \text{ cm}}{1 \text{ in}}$$

Step 2 Determine which conversion factor to use. You are converting from cm to in. Use the conversion factor with cm on the bottom.

$$\frac{1 \text{ in}}{2.54 \text{ cm}}$$

Step 3 Multiply the starting quantity and units by the conversion factor. Cancel the starting units with the units in the denominator. Round your answer based on the number of significant figures in the conversion factor.

$$100 \text{ cm} \times \frac{1 \text{ in}}{2.54 \text{ cm}} = 39.37 \text{ in}$$

The meterstick is 39.4 in long.

Practice Problem A book has a mass of 5 lbs. What is the mass of the book in kg?

Practice Problem Use the equivalent for in and cm (1 in = 2.54 cm) to show how $1 \text{ in}^3 = 16.39 \text{ cm}^3$.

Precision and Significant Digits

When you make a measurement, the value you record depends on the precision of the measuring instrument. This precision is represented by the number of significant digits recorded in the measurement. When counting the number of significant digits, all digits are counted except zeros at the end of a number with no decimal point such as 2,050, and zeros at the beginning of a decimal such as 0.03020. When adding or subtracting numbers with different precision, round the answer to the smallest number of decimal places of any number in the sum or difference. When multiplying or dividing, the answer is rounded to the smallest number of significant digits of any number being multiplied or divided.

Example The lengths 5.28 and 5.2 are measured in meters. Find the sum of these lengths and record your answer using the correct number of significant digits.

Step 1 Find the sum.

5.28 m	2 digits after the decimal
+ 5.2 m	1 digit after the decimal
10.48 m	

Step 2 Round to one digit after the decimal because the least number of digits after the decimal of the numbers being added is 1.

The sum is 10.5 m.

Practice Problem How many significant digits are in the measurement 7,071,301 m? How many significant digits are in the measurement 0.003010 g?

Practice Problem Multiply 5.28 and 5.2 using the rule for multiplying and dividing. Record the answer using the correct number of significant digits.

Scientific Notation

Many times numbers used in science are very small or very large. Because these numbers are difficult to work with scientists use scientific notation. To write numbers in scientific notation, move the decimal point until only one non-zero digit remains on the left. Then count the number of places you moved the decimal point and use that number as a power of ten. For example, the average distance from the Sun to Mars is 227,800,000,000 m. In scientific notation, this distance is 2.278×10^{11} m. Because you moved the decimal point to the left, the number is a positive power of ten.

The mass of an electron is about 0.000 000 000 000 000 000 000 000 000 000 911 kg. Expressed in scientific notation, this mass is 9.11×10^{-31} kg. Because the decimal point was moved to the right, the number is a negative power of ten.

Example Earth is 149,600,000 km from the Sun. Express this in scientific notation.

Step 1 Move the decimal point until one non-zero digit remains on the left.
1.496 000 00

Step 2 Count the number of decimal places you have moved. In this case, eight.

Step 3 Show that number as a power of ten, 10^8.

The Earth is 1.496×10^8 km from the Sun.

Practice Problem How many significant digits are in 149,600,000 km? How many significant digits are in 1.496×10^8 km?

Practice Problem Parts used in a high performance car must be measured to 7×10^{-6} m. Express this number as a decimal.

Practice Problem A CD is spinning at 539 revolutions per minute. Express this number in scientific notation.

Make and Use Graphs

Data in tables can be displayed in a graph—a visual representation of data. Common graph types include line graphs, bar graphs, and circle graphs.

Line Graph A line graph shows a relationship between two variables that change continuously. The independent variable is changed and is plotted on the *x*-axis. The dependent variable is observed, and is plotted on the *y*-axis.

Example Draw a line graph of the data below from a cyclist in a long-distance race.

Table 5 Bicycle Race Data	
Time (h)	Distance (km)
0	0
1	8
2	16
3	24
4	32
5	40

Step 1 Determine the *x*-axis and *y*-axis variables. Time varies independently of distance and is plotted on the *x*-axis. Distance is dependent on time and is plotted on the *y*-axis.

Step 2 Determine the scale of each axis. The *x*-axis data ranges from 0 to 5. The *y*-axis data ranges from 0 to 40.

Step 3 Using graph paper, draw and label the axes. Include units in the labels.

Step 4 Draw a point at the intersection of the time value on the *x*-axis and corresponding distance value on the *y*-axis. Connect the points and label the graph with a title, as shown in **Figure 20.**

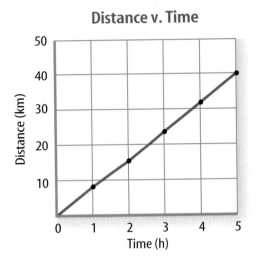

Figure 20 This line graph shows the relationship between distance and time during a bicycle ride.

Practice Problem A puppy's shoulder height is measured during the first year of her life. The following measurements were collected: (3 mo, 52 cm), (6 mo, 72 cm), (9 mo, 83 cm), (12 mo, 86 cm). Graph this data.

Find a Slope The slope of a straight line is the ratio of the vertical change, rise, to the horizontal change, run.

$$\text{Slope} = \frac{\text{vertical change (rise)}}{\text{horizontal change (run)}} = \frac{\text{change in } y}{\text{change in } x}$$

Example Find the slope of the graph in **Figure 20.**

Step 1 You know that the slope is the change in *y* divided by the change in *x*.
$$\text{Slope} = \frac{\text{change in } y}{\text{change in } x}$$

Step 2 Determine the data points you will be using. For a straight line, choose the two sets of points that are the farthest apart.
$$\text{Slope} = \frac{(40-0) \text{ km}}{(5-0) \text{ hr}}$$

Step 3 Find the change in *y* and *x*.
$$\text{Slope} = \frac{40 \text{ km}}{5 \text{h}}$$

Step 4 Divide the change in *y* by the change in *x*.
$$\text{Slope} = \frac{8 \text{ km}}{\text{h}}$$

The slope of the graph is 8 km/h.

Bar Graph To compare data that does not change continuously you might choose a bar graph. A bar graph uses bars to show the relationships between variables. The *x*-axis variable is divided into parts. The parts can be numbers such as years, or a category such as a type of animal. The *y*-axis is a number and increases continuously along the axis.

Example A recycling center collects 4.0 kg of aluminum on Monday, 1.0 kg on Wednesday, and 2.0 kg on Friday. Create a bar graph of this data.

Step 1 Select the *x*-axis and *y*-axis variables. The measured numbers (the masses of aluminum) should be placed on the *y*-axis. The variable divided into parts (collection days) is placed on the *x*-axis.

Step 2 Create a graph grid like you would for a line graph. Include labels and units.

Step 3 For each measured number, draw a vertical bar above the *x*-axis value up to the *y*-axis value. For the first data point, draw a vertical bar above Monday up to 4.0 kg.

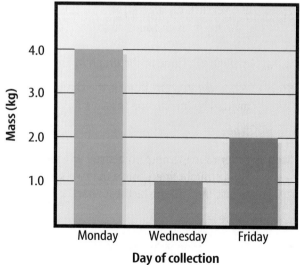

Aluminum Collected During Week

Practice Problem Draw a bar graph of the gases in air: 78% nitrogen, 21% oxygen, 1% other gases.

Circle Graph To display data as parts of a whole, you might use a circle graph. A circle graph is a circle divided into sections that represent the relative size of each piece of data. The entire circle represents 100%, half represents 50%, and so on.

Example Air is made up of 78% nitrogen, 21% oxygen, and 1% other gases. Display the composition of air in a circle graph.

Step 1 Multiply each percent by 360° and divide by 100 to find the angle of each section in the circle.

$$78\% \times \frac{360°}{100} = 280.8°$$

$$21\% \times \frac{360°}{100} = 75.6°$$

$$1\% \times \frac{360°}{100} = 3.6°$$

Step 2 Use a compass to draw a circle and to mark the center of the circle. Draw a straight line from the center to the edge of the circle.

Step 3 Use a protractor and the angles you calculated to divide the circle into parts. Place the center of the protractor over the center of the circle and line the base of the protractor over the straight line.

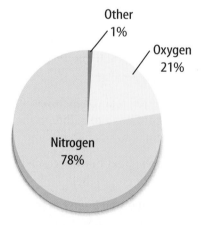

Other 1%
Oxygen 21%
Nitrogen 78%

Practice Problem Draw a circle graph to represent the amount of aluminum collected during the week shown in the bar graph to the left.

Physical Science Reference Tables

Standard Units

Symbol	Name	Quantity
m	meter	length
kg	kilogram	mass
Pa	pascal	pressure
K	kelvin	temperature
mol	mole	amount of a substance
J	joule	energy, work, quantity of heat
s	second	time
C	coulomb	electric charge
V	volt	electric potential
A	ampere	electric current
Ω	ohm	resistance

Physical Constants and Conversion Factors

Acceleration due to gravity	g	9.8 m/s/s or m/s^2
Avogadro's Number	N_A	6.02×10^{23} particles per mole
Electron charge	e	1.6×10^{-19} C
Electron rest mass	m_e	9.11×10^{-31} kg
Gravitation constant	G	6.67×10^{-11} N $\times$ m^2/kg^2
Mass-energy relationship		1 u (amu) $= 9.3 \times 10^2$ MeV
Speed of light in a vacuum	c	3.00×108 m/s
Speed of sound at STP		331 m/s
Standard Pressure		1 atmosphere
		101.3 kPa
		760 Torr or mmHg
		14.7 lb/in.^{2}

Wavelengths of Light in a Vacuum

Violet	$4.0 - 4.2 \times 10^{-7}$ m
Blue	$4.2 - 4.9 \times 10^{-7}$ m
Green	$4.9 - 5.7 \times 10^{-7}$ m
Yellow	$5.7 - 5.9 \times 10^{-7}$ m
Orange	$5.9 - 6.5 \times 10^{-7}$ m
Red	$6.5 - 7.0 \times 10^{-7}$ m

The Index of Refraction for Common Substances
($\lambda = 5.9 \times 10^{-7}$ m)

Air	1.00
Alcohol	1.36
Canada Balsam	1.53
Corn Oil	1.47
Diamond	2.42
Glass, Crown	1.52
Glass, Flint	1.61
Glycerol	1.47
Lucite	1.50
Quartz, Fused	1.46
Water	1.33

Heat Constants

	Specific Heat (average) (kJ/kg × °C) (J/g × °C)	Melting Point (°C)	Boiling Point (°C)	Heat of Fusion (kJ/kg) (J/g)	Heat of Vaporization (kJ/kg) (J/g)
Alcohol (ethyl)	2.43 (liq.)	−117	79	109	855
Aluminum	0.90 (sol.)	660	2467	396	10500
Ammonia	4.71 (liq.)	−78	−33	332	1370
Copper	0.39 (sol.)	1083	2567	205	4790
Iron	0.45 (sol.)	1535	2750	267	6290
Lead	0.13 (sol.)	328	1740	25	866
Mercury	0.14 (liq.)	−39	357	11	295
Platinum	0.13 (sol.)	1772	3827	101	229
Silver	0.24 (sol.)	962	2212	105	2370
Tungsten	0.13 (sol.)	3410	5660	192	4350
Water (solid)	2.05 (sol.)	0	–	334	–
Water (liquid)	4.18 (liq.)	–	100	–	–
Water (vapor)	2.01 (gas)	–	–	–	2260
Zinc	0.39 (sol.)	420	907	113	1770

PERIODIC TABLE OF THE ELEMENTS

Columns of elements are called groups. Elements in the same group have similar chemical properties.

	Gas
	Liquid
	Solid
	Synthetic

Element — Hydrogen
Atomic number — 1
Symbol — **H**
Atomic mass — 1.008

State of matter

The first three symbols tell you the state of matter of the element at room temperature. The fourth symbol identifies elements that are not present in significant amounts on Earth. Useful amounts are made synthetically.

	1	2		3	4	5	6	7	8	9
1	Hydrogen 1 **H** 1.008									
2	Lithium 3 **Li** 6.941	Beryllium 4 **Be** 9.012								
3	Sodium 11 **Na** 22.990	Magnesium 12 **Mg** 24.305								
4	Potassium 19 **K** 39.098	Calcium 20 **Ca** 40.078	Scandium 21 **Sc** 44.956	Titanium 22 **Ti** 47.867	Vanadium 23 **V** 50.942	Chromium 24 **Cr** 51.996	Manganese 25 **Mn** 54.938	Iron 26 **Fe** 55.845	Cobalt 27 **Co** 58.933
5	Rubidium 37 **Rb** 85.468	Strontium 38 **Sr** 87.62	Yttrium 39 **Y** 88.906	Zirconium 40 **Zr** 91.224	Niobium 41 **Nb** 92.906	Molybdenum 42 **Mo** 95.94	Technetium 43 **Tc** (98)	Ruthenium 44 **Ru** 101.07	Rhodium 45 **Rh** 102.906
6	Cesium 55 **Cs** 132.905	Barium 56 **Ba** 137.327	Lanthanum 57 **La** 138.906	Hafnium 72 **Hf** 178.49	Tantalum 73 **Ta** 180.948	Tungsten 74 **W** 183.84	Rhenium 75 **Re** 186.207	Osmium 76 **Os** 190.23	Iridium 77 **Ir** 192.217
7	Francium 87 **Fr** (223)	Radium 88 **Ra** (226)	Actinium 89 **Ac** (227)	Rutherfordium 104 **Rf** (261)	Dubnium 105 **Db** (262)	Seaborgium 106 **Sg** (266)	Bohrium 107 **Bh** (264)	Hassium 108 **Hs** (277)	Meitnerium 109 **Mt** (268)

The number in parentheses is the mass number of the longest-lived isotope for that element.

Rows of elements are called periods. Atomic number increases across a period.

The arrow shows where these elements would fit into the periodic table. They are moved to the bottom of the table to save space.

Lanthanide series	Cerium 58 **Ce** 140.116	Praseodymium 59 **Pr** 140.908	Neodymium 60 **Nd** 144.24	Promethium 61 **Pm** (145)	Samarium 62 **Sm** 150.36
Actinide series	Thorium 90 **Th** 232.038	Protactinium 91 **Pa** 231.036	Uranium 92 **U** 238.029	Neptunium 93 **Np** (237)	Plutonium 94 **Pu** (244)

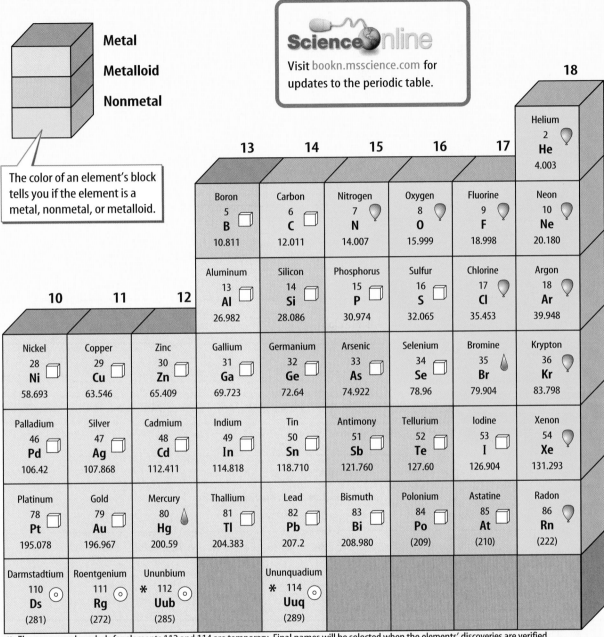

Metal

Metalloid

Nonmetal

The color of an element's block tells you if the element is a metal, nonmetal, or metalloid.

Science Online

Visit bookn.msscience.com for updates to the periodic table.

* The names and symbols for elements 112 and 114 are temporary. Final names will be selected when the elements' discoveries are verified.

Standard Units

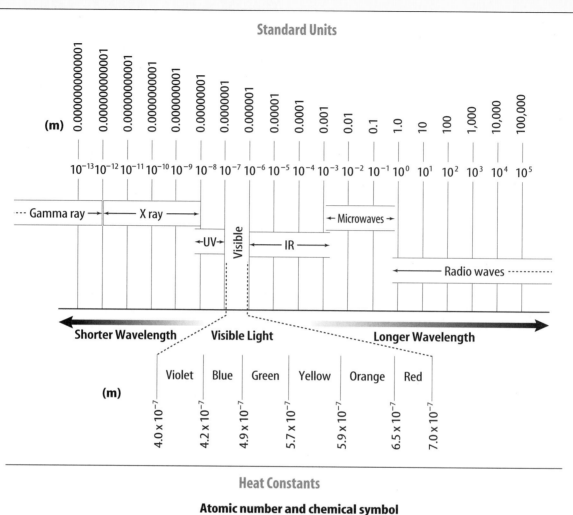

Heat Constants

Atomic number and chemical symbol

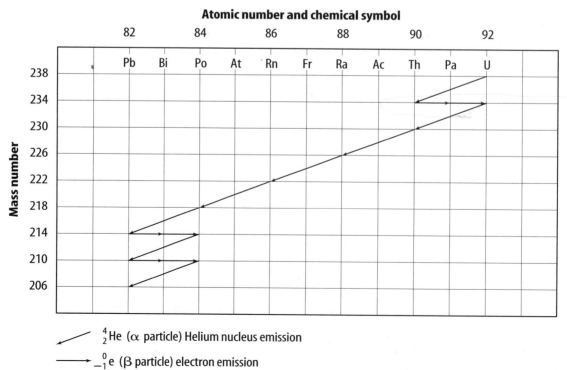

$^4_2$He (α particle) Helium nucleus emission

$^0_{-1}$e (β particle) electron emission

Glossary/Glosario

Cómo usar el glosario en español:
1. Busca el término en inglés que desees encontrar.
2. El término en español, junto con la definición, se encuentran en la columna de la derecha.

Pronunciation Key

Use the following key to help you sound out words in the glossary.

a . . . . back (BAK)		ew . . . . food (FEWD)	
ay . . . . day (DAY)		yoo . . . . pure (PYOOR)	
ah . . . . father (FAH thur)		yew . . . . few (FYEW)	
ow . . . . flower (FLOW ur)		uh . . . . comma (CAH muh)	
ar . . . . car (CAR)		u (+ con) . . . . . . rub (RUB)	
e . . . . less (LES)		sh . . . . shelf (SHELF)	
ee . . . . leaf (LEEF)		ch . . . . nature (NAY chur)	
ih . . . . trip (TRIHP)		g . . . . gift (GIHFT)	
i (i + con + e) . . idea (i DEE uh)		j . . . . gem (JEM)	
oh . . . . go (GOH)		ing . . . . sing (SING)	
aw . . . . soft (SAWFT)		zh . . . . vision (VIH zhun)	
or . . . . orbit (OR buht)		k . . . . cake (KAYK)	
oy . . . . coin (COYN)		s . . . . seed, cent (SEED, SENT)	
oo . . . . foot (FOOT)		z . . . . zone, raise (ZOHN, RAYZ)	

English — A — Español

alternating current (AC): electric current that changes its direction repeatedly. (p. 50)

analog signal: a electronic signal that carries information and varies smoothly with time. (p. 66)

aurora: light display that occurs when charged particles trapped in the magnetosphere collide with Earth's atmosphere above the poles. (p. 49)

corriente alterna (CA): corriente eléctrica que cambia de dirección repetidamente. (p. 50)

señal analógica: señal electrónica que conduce información y varía de manera uniforme con el tiempo. (p. 66)

aurora: despliegue de luz que se produce cuando partículas cargadas atrapadas en la magnetosfera chocan contra la atmósfera terrestre por encima de los polos. (p. 49)

B

binary system: number system consisting of two digits, 0 and 1, that can be used by devices such as computers to store or use information. (p. 74)

sistema binario: sistema numérico que consiste en dos dígitos, 0 y 1, que se puede usar con dispositivos como las computadoras para almacenar o usar información. (p. 74)

C

circuit: closed conducting loop in which electric current can flow continually. (p. 15)

computer software: any list of instructions for a computer to follow that is stored in the computer's memory. (p. 77)

circuito: circuito conductor cerrado en el cual la energía puede fluir continuamente. (p. 15)

software para computadoras: cualquier lista de instrucciones que debe realizar una computadora y que se almacena en la memoria de ésta. (p. 77)

Glossary/Glosario

conductor: material in which electrons can move easily. (p. 12)

conductor: material en el cual los electrones se pueden mover fácilmente. (p. 12)

D

digital signal: electronic signal that varies information that does not vary smoothly with time, but changes in steps between certain values, and can be represented by a series of numbers. (p. 67)

diode: a solid-state component made from two layers of semiconductor material that allows electric current to flow in only one direction and is commonly used to change alternating current to direct current. (p. 70)

direct current (DC): electric current that flows only in one direction. (p. 51)

señal digital: señal electrónica que varía aquella información que no varía de manera uniforme con el tiempo, pero que cambia por grados entre ciertos valores y que puede ser representada por una serie de números. (p. 67)

diodo: componente de estado sólido conformado por dos capas de material semiconductor que permite el flujo de corriente eléctrica en una sola dirección y que comúnmente se utiliza para cambiar la corriente alterna a corriente directa. (p. 70)

corriente directa (CD): corriente eléctrica que fluye solamente en una dirección. (p. 51)

E

electric current: the flow of electric charge, measured in amperes (A). (p. 15)

electric discharge: rapid movement of excess charge from one place to another. (p. 13)

electric field: surrounds every electric charge and exerts forces on other electric charges. (p. 11)

electric force: attractive or repulsive force exerted by all charged objects on each other. (p. 11)

electric power: rate at which electrical energy is converted into other forms of energy, measured in watts (W) or kilowatts (kW). (p. 24)

electromagnet: magnet created by wrapping a current-carrying wire around an iron core. (p. 45)

electronic signal: a changing electric current that is used to carry information; can be analog or digital. (p. 66)

corriente eléctrica: flujo de carga eléctrica, el cual se mide en amperios (A). (p. 15)

descarga eléctrica: movimiento rápido de carga excesiva de un lugar a otro. (p. 13)

campo eléctrico: campo que rodea a todas las cargas eléctricas y que ejerce fuerzas sobre otras cargas eléctricas. (p. 11)

fuerza eléctrica: fuerza de atracción o de repulsión que ejercen todos los objetos cargados entre ellos mismos. (p. 11)

potencia eléctrica: tasa a la cual la energía eléctrica se convierte en otras formas de energía, la cual se mide en vatios (W) o en kilovatios (kW). (p. 24)

electroimán: imán que se crea al enrollar un cable transportador de corriente alrededor de un centro de hierro. (p. 45)

señal electrónica: corriente eléctrica dinámica que se usa para conducir información; puede ser analógica o digital. (p. 66)

G

generator: device that uses a magnetic field to turn kinetic energy into electrical energy. (p. 50)

generador: dispositivo que utiliza un campo magnético para convertir energía cinética en energía eléctrica. (p. 50)

I

insulator: material in which electrons cannot move easily. (p. 12)

integrated circuit: circuit that can contain millions of interconnected transistors and diodes imprinted on a single small chip of semiconductor material. (p. 71)

ion: atom that is positively or negatively charged. (p. 8)

aislante: material en el cual los electrones no se pueden mover fácilmente.(p. 12)

circuito integrado: circuito que puede contener millones de transistores y diodos interconectados y fijados en un solo chip de tamaño reducido y hecho de material semiconductor. (p. 71)

ion: átomo cargado positiva o negativamente. (p. 8)

M

magnetic domain: group of atoms whose fields point in the same direction. (p. 40)

magnetic field: surrounds a magnet and exerts a magnetic force on other magnets. (p. 39)

magnetosphere: region of space affected by Earth's magnetic field. (p. 41)

microprocessor: integrated circuit that controls the flow of information between different parts of the computer; also called the central processing unit or CPU. (p. 79)

motor: device that transforms electrical energy into kinetic energy. (p. 48)

dominio magnético: grupo de átomos cuyos campos apuntan en la misma dirección. (p. 40)

campo magnético: campo que rodea a un imán y ejerce fuerza magnética sobre otros imanes. (p. 39)

magnetosfera: región del espacio afectada por el campo magnético de la Tierra. (p. 41)

microprocesador: circuito integrado que controla el flujo de información entre diferentes partes de una computadora; también se lo denomina la unidad central de procesamiento o CPU. (p. 79)

motor: dispositivo que transforma energía eléctrica en energía cinética. (p. 48)

O

Ohm's law: states that the current in a circuit equals the voltage divided by the resistance in the circuit. (p. 21)

ley de Ohm: establece que la corriente en un circuito es igual al voltaje dividido por la resistencia en el circuito. (p. 21)

P

parallel circuit: circuit that has more than one path for electric current to follow. (p. 23)

circuito paralelo: circuito en el cual la corriente eléctrica puede seguir más de una trayectoria. (p. 23)

R

random-access memory (RAM): temporary electronic memory within a computer. (p. 76)

read-only memory (ROM): electronic memory that is permanently stored within a computer. (p. 76)

memoria de acceso aleatorio (RAM): memoria electrónica temporal dentro de una computadora. (p. 76)

memoria de sólo lectura (ROM): memoria electrónica almacenada permanentemente dentro de una computadora. (p. 76)

Glossary/Glosario

resistance: a measure of how difficult it is for electrons to flow in a material; unit is the ohm (Ω). (p. 18)

resistencia: medida de la dificultad que tienen los electrones para fluir en un material; se mide en ohmios (Ω). (p. 18)

S

semiconductor: element, such as silicon, that is a poorer electrical conductor that a metal, but a better conductor than a nonmetal, and whose electrical conductivity can be changed by adding impurities. (p. 69)

series circuit: circuit that has only one path for electric current to follow. (p. 22)

static charge: imbalance of electric charge on an object. (p. 9)

semiconductor: elemento, como el silicio, que no es tan buen conductor de electricidad como un metal, pero que es mejor conductor que un no metal y cuya conductividad eléctrica puede ser modificada al añadirle impurezas. (p. 69)

circuito en serie: circuito en el cual la corriente eléctrica sólo puede seguir una trayectoria. (p. 22)

carga estática: desequilibrio de la carga eléctrica en un objeto. (p. 9)

T

transformer: device used to increase or decrease the voltage of an alternating current. (p. 52)

transistor: a solid-state component made from three layers of semiconductor material that can amplify the strength of an electric signal or act as an electronic switch. (p. 71)

transformador: dispositivo utilizado para aumentar o disminuir el voltaje de una corriente alterna. (p. 52)

transistor: componente de estado sólido formado por tres capas de material semiconductor que puede amplificar la fuerza de una señal eléctrica o actuar a manera de interruptor electrónico. (p. 71)

V

voltage: a measure of the amount of electrical potential energy an electron flowing in a circuit can gain; measured in volts (V). (p. 16)

voltaje: medida de la cantidad de energía eléctrica potencial que puede adquirir un electrón que fluye en un circuito; se mide en voltios (V). (p. 16)

Magnification Key: Magnifications listed are the magnifications at which images were originally photographed.
LM–Light Microscope
SEM–Scanning Electron Microscope
TEM–Transmission Electron Microscope

Acknowledgments: Glencoe would like to acknowledge the artists and agencies who participated in illustrating this program: Absolute Science Illustration; Andrew Evansen; Argosy; Articulate Graphics; Craig Attebery, represented by Frank & Jeff Lavaty; CHK America; John Edwards and Associates; Gagliano Graphics; Pedro Julio Gonzalez, represented by Melissa Turk & The Artist Network; Robert Hynes, represented by Mendola Ltd.; Morgan Cain & Associates; JTH Illustration; Laurie O'Keefe; Matthew Pippin, represented by Beranbaum Artist's Representative; Precision Graphics; Publisher's Art; Rolin Graphics, Inc.; Wendy Smith, represented by Melissa Turk & The Artist Network; Kevin Torline, represented by Berendsen and Associates, Inc.; WILDlife ART; Phil Wilson, represented by Cliff Knecht Artist Representative; Zoo Botanica.

Photo Credits

Cover PhotoDisc; **i ii** PhotoDisc; **iv** (bkgd)John Evans, (inset)PhotoDisc; **v** (t)PhotoDisc, (b)John Evans; **vi** (l)John Evans, (r)Geoff Butler; **vii** (l)John Evans, (r)PhotoDisc; **viii** PhotoDisc; **ix** Aaron Haupt Photography; **x** V.C.L./Getty Images; **xi** (l)Thomas Brummett/PhotoDisc, (r)courtesy IBM/Florida State University; **xii** (t)Richard Hutchings, (b)James Leynse/CORBIS; **1** Norbert Schafer/CORBIS; **2** (t)CORBIS, (b)SuperStock; **3** (tr)Argosy, (br)Ace Photo Agency/PhotoTake NYC; **4** Lester Lefkowitz/The Stock Market; **5** (t)AP/Wide World Photos, (b)Don Farrall/ PhotoDisc; **7–8** V.C.L./Getty Images; **9** (t)Richard Hutchings, (b)KS Studios; **12** Royalty Free/CORBIS; **14** J. Tinning/Photo Researchers; **17** Gary Rhijnsburger/Masterfile; **22** Doug Martin; **23** (t)Doug Martin, (b)Geoff Butler; **25** Bonnie Freer/Photo Researchers; **27** Matt Meadows; **28 29** Richard Hutchings; **30** (bkgd)Tom & Pat Leeson/Photo Researchers, (inset)William Munoz/Photo Researchers; **34** J. Tinning/ Photo Researchers; **35** Doug Martin; **36–37** James Leynse/ CORBIS; **39** Richard Megna/Fundamental Photographs; **40** Amanita Pictures; **43** John Evans; **44** Amanita Pictures; **45** (l)Kodansha, (c)Manfred Kage/Peter Arnold, Inc., (r)Doug Martin; **49** Bjorn Backe/Papilio/CORBIS; **51** Norbert Schafer/ The Stock Market/CORBIS; **53** AT&T Bell Labs/Science Photo Library/Photo Researchers; **54** (t)Science Photo Library/Photo Researchers, (c)Fermilab/Science Photo Library/Photo Researchers, (b)SuperStock; **55** PhotoDisc; **56** (t)file photo, (b)Aaron Haupt; **57** Aaron Haupt; **58** John MacDonald; **59** (l)SIU/Peter Arnold, Inc., (r)Latent Image; **63** John Evans; **64–65** Andrew Syred/Science Photo Library/ Photo Researchers; **66** Willie L. Hill, Jr./Stock Boston; **67** (l)Icon Images, (c)Russ Lappa, (r)Doug Martin; **69** CMCD/PhotoDisc; **70** Amanita Pictures; **71** (t)Amanita Pictures, (b)Charles Falco/Photo Researchers; **72** Charles Falco/Photo Researchers; **73** (l)Bettmann/CORBIS, (r)Icon Images; **76** (t)courtesy IBM/Florida State University, (b)Andrew Syred/Science Photo Library/Photo Researchers; **79** file photo; **80** Thomas Brummett/PhotoDisc; **82** (l)Dr. Dennis Kunkel/PhotoTake, NYC, (r)Aaron Haupt; **83** Timothy Fuller; **84** David Young-Wolff/PhotoEdit, Inc.; **85** Frank Cezus; **86** Tek Images/Science Photo Library/Photo Researchers; **87** (tr)Amanita Pictures, (l)Aaron Haupt, (br)Keith Brofsky/PhotoDisc; **91** Thomas Brummett/ PhotoDisc; **92** PhotoDisc; **94** Tom Pantages; **96** Michell D. Bridwell/PhotoEdit, Inc.; **99** (t)Mark Burnett, (b)Dominic Oldershaw; **100** StudiOhio; **101** Timothy Fuller; **102** Aaron Haupt; **104** KS Studios; **105** Matt Meadows; **106** Icon Images; **108** Amanita Pictures; **109** Bob Daemmrich; **111** Davis Barber/PhotoEdit, Inc.

PERIODIC TABLE OF THE ELEMENTS

Columns of elements are called groups. Elements in the same group have similar chemical properties.

Gas
Liquid
Solid
Synthetic

Element —— Hydrogen
Atomic number —— 1
Symbol —— **H**
Atomic mass —— 1.008
State of matter

The first three symbols tell you the state of matter of the element at room temperature. The fourth symbol identifies elements that are not present in significant amounts on Earth. Useful amounts are made synthetically.

Group	1	2	3	4	5	6	7	8	9
1	Hydrogen 1 **H** 1.008								
2	Lithium 3 **Li** 6.941	Beryllium 4 **Be** 9.012							
3	Sodium 11 **Na** 22.990	Magnesium 12 **Mg** 24.305							
4	Potassium 19 **K** 39.098	Calcium 20 **Ca** 40.078	Scandium 21 **Sc** 44.956	Titanium 22 **Ti** 47.867	Vanadium 23 **V** 50.942	Chromium 24 **Cr** 51.996	Manganese 25 **Mn** 54.938	Iron 26 **Fe** 55.845	Cobalt 27 **Co** 58.933
5	Rubidium 37 **Rb** 85.468	Strontium 38 **Sr** 87.62	Yttrium 39 **Y** 88.906	Zirconium 40 **Zr** 91.224	Niobium 41 **Nb** 92.906	Molybdenum 42 **Mo** 95.94	Technetium 43 **Tc** (98)	Ruthenium 44 **Ru** 101.07	Rhodium 45 **Rh** 102.906
6	Cesium 55 **Cs** 132.905	Barium 56 **Ba** 137.327	Lanthanum 57 **La** 138.906	Hafnium 72 **Hf** 178.49	Tantalum 73 **Ta** 180.948	Tungsten 74 **W** 183.84	Rhenium 75 **Re** 186.207	Osmium 76 **Os** 190.23	Iridium 77 **Ir** 192.217
7	Francium 87 **Fr** (223)	Radium 88 **Ra** (226)	Actinium 89 **Ac** (227)	Rutherfordium 104 **Rf** (261)	Dubnium 105 **Db** (262)	Seaborgium 106 **Sg** (266)	Bohrium 107 **Bh** (264)	Hassium 108 **Hs** (277)	Meitnerium 109 **Mt** (268)

The number in parentheses is the mass number of the longest-lived isotope for that element.

Rows of elements are called periods. Atomic number increases across a period.

The arrow shows where these elements would fit into the periodic table. They are moved to the bottom of the table to save space.

Lanthanide series	Cerium 58 **Ce** 140.116	Praseodymium 59 **Pr** 140.908	Neodymium 60 **Nd** 144.24	Promethium 61 **Pm** (145)	Samarium 62 **Sm** 150.36
Actinide series	Thorium 90 **Th** 232.038	Protactinium 91 **Pa** 231.036	Uranium 92 **U** 238.029	Neptunium 93 **Np** (237)	Plutonium 94 **Pu** (244)